OPERATION
CHASTISE

Also by Max Hastings

REPORTAGE

America 1968: The Fire this Time
Ulster 1969: The Struggle for Civil Rights in Northern Ireland
The Battle for the Falklands (with Simon Jenkins)

BIOGRAPHY

Montrose: The King's Champion
Yoni: Hero of Entebbe

AUTOBIOGRAPHY

Did You Really Shoot the Television?
Going to the Wars
Editor

HISTORY

Bomber Command
The Battle of Britain (with Len Deighton)
Das Reich
Overlord: D-Day and the Battle for Normandy
Victory in Europe
The Korean War
Warriors: Extraordinary Tales from the Battlefield
Armageddon: The Battle for Germany 1944–45
Nemesis: The Battle for Japan 1944–45
Finest Years: Churchill as Warlord 1940–45
All Hell Let Loose: The World at War 1939–45
Catastrophe: Europe Goes to War 1914
The Secret War: Spies, Codes and Guerrillas 1939–1945
Vietnam: An Epic Tragedy 1945–1975

COUNTRYSIDE WRITING

Outside Days
Scattered Shots
Country Fair

ANTHOLOGY (EDITED)

The Oxford Book of Military Anecdotes

MAX HASTINGS

OPERATION CHASTISE

*THE RAF's MOST BRILLIANT ATTACK
OF WORLD WAR II*

HARPER

An Imprint of HarperCollinsPublishers

HarperCollins books may be purchased for educational, business, or sales promotional use. For information, please email the Special Markets Department at SPsales@harpercollins.com.

Originally published as *Chastise* in Great Britain in 2019 by William Collins, an imprint of HarperCollins Publishers.

FIRST U.S. EDITION

Library of Congress Cataloging-in-Publication Data has been applied for.

ISBN 978-0-06-295363-6

20 21 22 23 24   LSC   10 9 8 7 6 5 4 3 2 1

In memory of the aircrew who achieved the
almost impossible on the night of 16/17 May 1943;
and of the men, women and children on
both sides who perished

Contents

Illustrations

(National Portrait Gallery, London/Howard Coster);
Arthur Collins; Air Marshal Sir John Linnell. (Imperial
War Museum/CM 5259); Gp. Capt. F. W. Winterbotham.
(Barry James Gilmore/Fairfax Media/Getty Images)

Air Chief Marshal Sir Arthur Harris. (Imperial War
Museum/CH 13020)

Harris with his wife Jill and daughter Jackie. (Leonard
McCombe/Picture Post/Getty Images)

Avro Lancaster ED932 (G-George) being flown by Guy
Gibson during tests at Reculver. (Imperial War Museum/
FLM 2343)

An Upkeep bounces onto the seafront during tests at
Reculver. (Imperial War Museum/FLM 2343)

Tests at Reculver. (Imperial War Museum/FLM 2360; FLM
2362; FLM 2363)

Melvin Young in the victorious 1938 Oxford Boat Race crew.
(The Times/News Licensing)

Young's miraculous October 1940 Atlantic rescue. (Trinity
College Oxford, with help from Arthur G. Thorning)

Young with his rescuer's captain. (Trinity College Oxford,
with help from Arthur G. Thorning)

Melvin and Priscilla Young. (Trinity College Oxford, with
help from Arthur G. Thorning)

John 'Hoppy' Hopgood. (Hopgood family)

Hopgood with his sister Marna and his mother. (Hopgood
family)

Hopgood with Marna. (Hopgood family)

Aircrew of 106 Squadron. (Imperial War Museum/HU
91941)

Henry Maudslay at Eton. (Maudslay family)

Maudslay as squadron leader. (Lincolnshire County Council
Archives)

Guy Gibson and his wife Eve. (Trinity Mirror/Mirrorpix/ Alamy)

Gibson in 1944. (Imperial War Museum/CH 13618)

Gibson with Nigger. (Associated Newspapers/Rota/ Shutterstock)

Gibson, Spafford, Hutchison, Deering and Taetum at Scampton. (CNP Collection/Alamy)

Bill Astell and his family on a pre-war outing in Derbyshire. (Ray Hepner)

Gibson with Dave Maltby. (CNP Collection/Alamy)

Australian crew members on leave in London after the raid. (Imperial War Museum/CH 9942)

F/Lt. Bill Astell. (Lincolnshire County Council Archives)

F/Lt. Joe McCarthy and his crew. (CNP Collection/Alamy)

P/O John Fraser at his wedding, a week before Chastise. (Cavendish Press)

F/Lt. David Shannon. (Central Press/Hulton Archive/ Getty)

F/Lt. Les Munro.

The aircrew who took part in Operation Chastise. (Imperial War Museum/CH 11049)

Lancaster taking off from Scampton for Chastise. (Imperial War Museum/CH 18006)

Wreck of a 617 Squadron Lancaster on a Dutch beach. (The National Archives/AIR 20/4367)

Harris and Cochrane at the debriefing of Gibson's crew. (CNP Collection/Alamy)

WAAF intelligence officer Fay Gillon with survivors of the raid, including Maltby, Munro, Trevor-Roper and Shannon. (Fay Gillon)

King George VI at Scampton, with Gibson and Whitworth. (CNP Collection/Alamy)

Boffin/bofin/n.*chiefly Br. informal*, a scientific expert; *esp.* one involved in technological research [origin unknown]

Longman's Dictionary of the English Language

'It is proposed to use this weapon … against a large dam in Germany which, if breached, will have serious consequences in the neighbouring industrial area … The operation … will not, it is thought, prove particularly dangerous, but it will undoubtedly require skilled crews … Some training will no doubt be necessary.'

Air Vice-Marshal Robert Oxland, Bomber Command HQ, to Air Vice-Marshal Ralph Cochrane, AOC 5 Group, on 17 March 1943

'One thing,' said Dim, 'if we do go and attack … one of us might possibly get a posthumous VC.' 'Who wants that?' said Taffy. 'Not me,' said one of the boys. 'All I want is a Peace and Victory Medal.' Most of us agreed.

Guy Gibson, *Enemy Coast Ahead*

After Hollywood mogul Daryl Zanuck was shown the movie *The Dam Busters* in 1955, he demanded disbelievingly, 'Is that a true story?' Yes, he was told. 'Then why doesn't it say so?'

RAF Ranks and Army Equivalents

Marshal of the RAF – Field-Marshal
Air Chief Marshal (ACM) – General
Air Marshal (AM) – Lieutenant-General
Air Vice-Marshal (AVM) – Major-General
Air Commodore (A/C) – Brigadier
Group-Captain (Gp. Capt.) – Colonel
Wing-Commander (W/Cdr.) – Lieutenant-Colonel
Squadron-Leader (S/Ldr.) – Major
Flight-Lieutenant (F/Lt.) – Captain
Flying Officer (F/O) – Lieutenant
Pilot Officer (P/O) – Subaltern
Flight-Sergeant (F/Sgt.) – Warrant Officer
Sergeant (Sgt.) – Sergeant
Corporal (Cpl.) – Corporal
Leading Aircraftman (LAC) – Lance-Corporal
Aircraftman (AC) – Private
Air Officer Commanding (AOC)

Ranks attributed to personnel mentioned in the text are those held at the time of incidents or conversations described.

Abbreviations Used in the Text

AOC – Air Officer Commanding

ATS – Auxiliary Territorial Service; women's branch of the army

CAS – Chief of the Air Staff; head of the RAF

C-in-C – Commander-in-Chief

CO – Squadron commanding officer

Gee – Electronic navigation aid, detecting a grid of radio signals transmitted from the UK, fitted to all Bomber Command aircraft but jammed by the Germans over continental Europe

HCU – Heavy Conversion Unit

IFF – Identification Friend or Foe: electronic radar-pulse identification device fitted to all British aircraft

MAP – Ministry of Aircraft Production

MEW – Ministry of Economic Warfare

OTU – Operational Training Unit

RAAF – Royal Australian Air Force

RAFVR – Royal Air Force Volunteer Reserve

RCAF – Royal Canadian Air Force

RNZAF – Royal New Zealand Air Force

SASO – Senior Air Staff Officer; comparable to an army or divisional commander's chief of staff

USAAF – United States Army Air Force
WAAF – Women's Auxiliary Air Force; thus a woman
serving at an RAF station would be described as a 'Waaf'
w/op – Wireless-operator

Narrative of operations uses a twenty-four-hour clock, while the twelve-hour civilian clock is used for other timings.

Bomber Command in February 1943 comprised around two thousand aircraft including trainers – the number varied daily, and significantly fewer were immediately serviceable – of which six hundred were 'heavies'. Each of seven operational Groups was commanded by an air vice-marshal, and contained variously five to ten squadrons. A squadron was composed of eighteen to twenty-four aircraft, confusingly led by a wing-commander, and subdivided into two or three flights, each commanded by a squadron-leader.

Introduction

There were dams; a dog with an embarrassing name; a movie; a march composed by Eric Coates. These memories of Operation Chastise, the 'bouncing bomb' attack which burst open north-western Germany's Möhne and Eder reservoirs on the night of 16/17 May 1943, cling to the consciousness of millions of people of all ages, both sexes and many nations, who may know little else about the Second World War. Wing-Commander Guy Gibson's biographer Richard Morris has written: 'The story of 617 Squadron's breaching of the dams has joined that group of historically-based tales – like King Arthur, or Robin Hood – which defy all efforts at scholarly revision.'

Much that we think we know is wrong. Those of us who read Paul Brickhill's 1951 book, then watched its 1955 screen progeny, the most popular British war film of all time, should blush to remember that we embraced *The Dam Busters* with special enthusiasm because the raid seemed victimless, save for the fifty-three dead among the gallant young men who carried it out. In truth, however, something approaching 1,400 people – almost all civilians and more than half French, Polish, Russian and Ukrainian mostly female slaves of Hitler – perished in the *Möhnekatastrophe*, as modern Germans call it;

more than in any previous RAF attack on the Reich. That tragic outcome deserves emphasis, alongside our awe at 617 Squadron's achievement. It is fascinating that Guy Gibson afterwards reflected uneasily about this, as his superiors never did, writing in 1944: 'The fact that people ... might drown had not occurred to us. But we hoped that the dam wardens would warn those living below in time, even if they were Germans. No one likes mass slaughter and we did not like being the authors of it. Besides, it brought us in line with Himmler and his boys.'

This book represents an emotional journey from my own childhood; from the day when, at boarding school, I first thrilled to Richard Todd's portrayal of the twenty-four-year-old Gibson, who led 617 Squadron on that fateful May night. Many legendary feats of courage have been performed by warriors who clung to some bleeding piece of earth: the Three Hundred at Thermopylae; Horatius on the bridge before Rome; the Guards' defence of Hougoumont at Waterloo; Joshua Lawrence Chamberlain's 20th Maine on Little Round Top at Gettysburg; C Company of the 24th Foot holding Rorke's Drift.

By contrast, the deed attempted in May 1943 by 130 British, Canadian, and Australian airmen, together with a single American and two New Zealanders, required qualities of a different order. Almost all were of an age with modern gap-year adolescents, or students at university. They embarked in cold blood on a mission that many recognised was likely to kill them, and that would require exceptional courage, skill and luck to succeed. They lifted their big, clumsy bombers from the tranquillity of a summer evening in the midst of the Lincolnshire countryside, barely four decades after the Wright brothers initiated heavier-than-air flight. For two and a half hours they raced through the moonlit sky towards Germany, at a height that made power cables as deadly a menace as anti-aircraft fire. They then attacked Hitler's dams, flying straight and level at 220 mph, much lower than the treetops and less than a cricket pitch's length from the lakes below, to unleash revolutionary four-and-a-half-ton weapons created by the brilliance and persistence of Barnes Wallis, a largely self-taught engineer. Half of 617's aircraft which got as far as Germany failed to return, but two of the biggest man-made structures in the world collapsed into mud and rubble, releasing hundreds of millions of tons of water upon the Reich.

The Allied bomber offensive has become one of the most controversial aspects of the Second World War. Some critics, not all of them German or Japanese, denounce the Western Allied assaults upon cities and their inhabitants as a war crime. The 1945 fire-bombing campaign by American B-29 Superfortresses killed far more people in Tokyo and other Japanese cities than did the atomic bombs later dropped on Hiroshima and Nagasaki. The concept of air bombardment of civilians causes many twenty-first-century people discomfort, indeed repugnance. Contrarily, it is a source of bitterness to some descendants of the RAF's wartime bomber crews that the public prefers to lavish legacy adulation on the Spitfire and Hurricane pilots of the Battle of Britain – defenders – than on their comrades the attackers, who bombed Germany at the cost of enduring losses much greater than those of Fighter Command. Australian Dave Shannon of 617 Squadron denounced in old age 'sanctimonious, hypocritical and grovelling criticism about things that were done in a total war'.

Where, in all this, does the saga of the dambusters rightfully belong? The fliers contrived a feat that caused all the world to wonder – the Allied nations with pride, the German people and their leaders with horror and apprehension. Rumours swept the Reich that thirty thousand victims had perished beneath the floods. Though the Lancaster crews were drawn from the RAF's Bomber Command, the force that nightly rained fire and destruction on Germany's cities, few even among its critics failed to perceive a nobility about the bravery displayed that night. In the spring of 1943, after nearly four years of austerity, unpalatable food, family separations and spasmodic terrors, only lately ameliorated by a thin gruel of successes, the British people were weary. The dams raid lifted their spirits, revived flagging confidence in their own nation's powers, as had few events since the desert victory at El Alamein six months earlier. We shall discuss below its effects on the Nazi war machine, which RAF planners aspired to cripple.

I was born at the end of 1945, and thus was five when Paul Brickhill's best-selling account of Chastise was published, nine when the film was released. Both book and movie made a profound impression. I memorised the names of almost every one of 617's pilots; assembled and painted plastic models of the Avro Lancasters they flew; became intimately familiar with *Enemy Coast Ahead*, Gibson's posthumously published memoir. As an adult, I began to study wars, first as a correspondent and eyewitness in faraway places, then as an author of books. Although my ideas became much more nuanced than those of childhood, I was well served by familiarity with a host of World War II memoirs and histories.

In 1977 I was commissioned to write *Bomber Command*, a study of the British strategic offensive. In those days, thousands of former aircrew were still alive, together with some commanders. I interviewed at length Air Chief Marshal Sir Arthur Harris, 1942–45 C-in-C of Bomber Command; Air Chief Marshal Sir Ralph Cochrane, ex-AOC of 5 Group, to which 617 Squadron belonged; former senior staff officers including Air Vice-Marshal Syd Bufton, director of bomber operations at the Air Ministry; Marshal of the RAF Lord Elworthy, who served as a pilot, station commander and staff officer at Bomber Command HQ; together with the inventor of the 'bouncing bomb', Sir Barnes Wallis, and wartime bombing adviser Lord Zuckerman. Among former 617 aircrew, I met Group-Captain Leonard Cheshire, VC, and Air Vice-Marshal Sir Harold 'Micky' Martin. Many of these disagreed profoundly with my conclusions, but their testimony was invaluable.

The RAF's Battle of Britain Flight flew me as a passenger from Farnborough to Coningsby in its only surviving Avro Lancaster, an unforgettable experience. I explored every crew position, and occupied the rear turret – albeit with most of my long back protruding through its sliding doors – while an accompanying Spitfire and Hurricane made passes, to give me a gunner's-eye view of an attacking fighter. As a war correspondent I saw more than a few aircraft shot down, and have myself dangled from a parachute, though happily not as a 'bailed out' airman. In 1994 I spent an airsick afternoon in the rear seat of an RAF Tornado of the latterday 617 Squadron, over Lincolnshire and the North Sea.

All these memories have informed my thoughts and stirred my imagination as I wrote this book. Among many previous accounts of Chastise, the 1982 groundbreaker was that of John Sweetman, who performed prodigies of research to transform and much enhance the picture created by Paul Brickhill. I cherish unstinting admiration for Richard Morris, and especially for his 1994 biography of Guy Gibson, which contributed much to Gibson's portrait in my own 2005 book *Warriors*. In Germany, Helmuth Euler has devoted most of his life to interviewing survivors of the breaching of the dams, as well as assembling images and documentation: I have made a free translation from some of his witnesses' testimony, in pursuit of colloquial English. In 2012, James Holland published an exhaustive new account of the raid.

Robert Owen, official historian of the 617 Squadron Association, possesses encyclopaedic knowledge, which he is generous enough to lend to other writers. Rob was a perfect companion on my 2018 visit to the dams, which enabled me to understand on the spot much that was previously obscure about the hazards facing the attackers. Charles Foster has recently published an invaluable new work of reference, *The Complete Dambusters*, providing images and details of all 133 aircrew who flew the raid. For my own narrative I have drawn heavily upon the researches of all the above writers. Richard Morris and Rob Owen, especially, have saved me from egregious errors. As in all my books, I seek to emphasise the human dimension and the 'big picture', making no attempt to match the admirable technical detail about Wallis's weapons, of which Sweetman and Holland display mastery.

An enigma overhangs the personalities of the men of 617 Squadron. Almost all were very young when they attacked the dams, and few survived the war. Records detail what they did; there is much less evidence, however, about what sort of people they were. With the notable exception of Gibson, their stories rely heavily upon adolescent correspondence and anecdotes. They were unformed in almost everything save having been trained for flight and devastation: many still thought it the best joke in the world to pull off a man's trousers after dinner. In describing them, an author cannot escape surmise and speculation. Much reported dialogue, especially relating to the hours of action over the dams, relies upon later personal memories, probably more reliable in spirit than wording. On such a matter as – for instance – the sporadic affair between Gibson and WAAF nurse Margaret North, historians depend on North's unsupported oral testimony to Richard Morris.

Since starting this book, I have been repeatedly asked whether it is an embarrassment to acknowledge the name of Gibson's dog, which became a wirelessed codeword for the breaching of the Möhne. A historian's answer must be: no more than the fact that our ancestors hanged sheep-stealers, executed military deserters and imprisoned homosexuals. They did and said things differently then. It would be grotesque to omit Nigger from a factual narrative merely because the word is rightly repugnant to twenty-first-century ears.

I have been moved to retell, and to reconsider, the Chastise story, in hopes of offering a new perspective which almost represents a paradox. I retain the awe of my childhood for the fliers who breached the Möhne and the Eder. In my seventies, I muse constantly upon the privilege of having attained old age, whereas the lives of most of those British, Canadian, Australian, New Zealand and American fliers became forfeit before they knew maturity, fatherhood or, in many cases, love or even sex.

Yet in the twenty-first century it also seems essential to confront – as many past British writers have been reluctant to confront – the enormity of the horror that the unthinking fliers unleashed upon a host of innocents. A Norwegian Resistance hero, Knut Lier-Hansen, wrote words in 1948 that linger in my mind whenever I compose narratives of conflict: 'Though wars can bring adventures which stir the heart, the true nature of war is composed of innumerable personal tragedies, of grief, waste and sacrifice, wholly evil and not redeemed by glory.' We shall consider below whether the extraordinary tale of Operation Chastise – its impact upon the Second World War set against its human consequences – is 'redeemed by glory'.

MAX HASTINGS
Chilton Foliat, West Berkshire, and Datai, Langkawi, Malaysia
May 2019

Prologue

Let us begin this story where he began it: in the cockpit of an Avro Lancaster heavy bomber, callsign G-George, forging through the darkness towards Germany on the night of 16 May 1943, amid the roar of four Rolls-Royce Merlin engines that drowned out conversation save over the intercom. 'The moon was full; everywhere its pleasant, watery haze spread over the peaceful English countryside, rendering it colourless. But there is not much colour in Lincolnshire anyway. The city of Lincoln was silent – that city which so many bomber boys know so well, a city full of homely people.' Guy Gibson's *Enemy Coast Ahead*, written in 1944, a few months before his death, is one of the great wartime warriors' memoirs, despite its cavalier attitude to facts and dates. Those who edited the typescript for publication after its author perished softened harshnesses: for instance, Gibson originally characterised Lincoln as 'full of dull, unimaginative people', perhaps because his own experiences were beyond their imaginations, and those of most of us.

The book reveals a sensitivity that few of the squadron commander's men recognised in him, together with a consciousness of his own mortality, derived from completion of an astounding seventy-two previous bomber operations, together with ninety-nine sorties as pilot of a night-fighter. He

describes the fate of an aircraft hit over Germany, plunging
steeply out of the sky for an interminable minute, 'then it is all
over and you hit the ground. Petrol flames come soaring up
into the sky, almost reaching to meet you as though to rocket
your soul to heaven.' He knew. Unlike some heroes who are
bereft of fear, Gibson anticipated his own almost certain
destiny. Only its hour remained to be fixed, and this May night
seemed more plausible than most.

He was leading 617 Squadron of the RAF's Bomber
Command to unleash upon the dams of north-western
Germany a revolutionary new weapon, requiring an attack at
extreme low level. Nineteen crews were committed to
Operation Chastise, and the eight proven in training to be
most proficient now accompanied Gibson himself towards the
Möhne. Off his port wingtip flew the dashing Australian
Harold 'Micky' Martin in P-Popsie, while a few yards to star-
board was 'Hoppy' Hopgood's M-Mother, its pilot a twenty-
one-year-old who still began his letters home 'Dear Mummy'.

Gibson again: 'We were off on a journey for which we had
long waited, a journey that had been carefully planned, carefully trained for' – only eight weeks, in truth, since inception,
but such a span represented an eternity to very young men,
crowding what should properly have been a lifetime's experience into a fraction of a natural span: Gibson considered en-
titling the later memoir of his career as a bomber pilot *Four
Years Lifetime* – 'a mission which was going to do a lot of good
if it succeeded'. The chiefs of the RAF had promised the aircrew
of 617 Squadron that breaching the dams would inflict damage
upon Germany's war industries greater than any previously
achieved by an air force.

Among the sharpest contrasts between the environment of
twentieth-century war and that of twenty-first-century peace

is colour. We live in a world of reds and whites, blues, silvers, oranges. Allied airmen bombing Europe in 1943 existed by day under sunshine and bright skies, then fought their battles in a universe that was darkened, shaded, shadowed, unless or until it erupted into flame. The undersides and flanks of night bombers were painted matt black; their upper surfaces disrupted foliage-greens and earth-browns. Once airborne, Gibson and his kin inhabited cramped, stunted workspaces, crowded with technology and control mechanisms, black or green save where paint had been worn away by human friction and hard usage to reveal streaks of dull metal.

He wrote: 'The pilot sits on the left on a raised, comfortably padded seat ... usually flies the thing with his left hand, re-setting the gyro and other instruments with his right, but most pilots use both hands when over enemy territory or when the going is tough. You have to be quite strong to fly a Lancaster. The instruments sit winking. On the Sperry panel, or the blind-flying panel as bomber pilots call it, now and then a red light, indicating that some mechanism needs adjusting, will suddenly flash on ... The pilot's eyes constantly perform a circle from the repeater to the Air Speed Indicator, from the ASI to the horizon, from the horizon to the moon, from the moon to what he can see on the ground and then back to the repeater. No wonder they are red-rimmed when he returns.'

Gibson himself had much to be red-rimmed about. Since 1940, he had been almost continuously making war. Two months earlier he completed a tour as 106 Squadron's commander, during which he flew its most hazardous opera-tions. He was now among the most decorated pilots in the RAF, holding two Distinguished Service Orders, two Distinguished Flying Crosses. In the seven weeks since he began to form 617 Squadron he had grappled with relentless administrative and personnel problems; directed specialised crew training; raced between Scampton, Weybridge and Reculver to discuss the dam-busting bomb with its creator and to witness the mixed fortunes of its trials. He had flown low-level tests by day and night; hastened to and from the Grantham headquarters of 5 Group for tactical conferences; delivered the most important briefings of his life; and now, taken off for Germany in the dying light of a lovely English spring Sunday.

Exhaustion most conspicuously manifested itself in an inflamed foot condition which caused pain on the ground, worse in the air. In a conversation with Gibson that morning, the station medical officer felt unable to prescribe medication, lest it impair the pilot's reflexes. Meanwhile Gibson's three-year marriage to an older showgirl had become a poor thing. His only relaxation in weeks had been a snatched trip to a Grantham 'flickhouse' with a WAAF girlfriend to see *Casablanca*. Like a host of young men of all the nations engaged in the Second World War, he had aged years beyond the twenty-four cited in RAF records.

That night of 16 May, he wrote of himself and John Pulford, the flight-engineer on the folding seat beside him in the Lancaster's 'glass house': 'two silent figures, young, unbearded, new to the world yet full of skill, full of pride in their squadron, determined to do a good job and bring the ship home. A silent scene, whose only incidental music is provided by the background hiss of air and the hearty roar of four Merlin engines.' He described Pulford as 'a Londoner, a sincere and plodding type'. He embraced Fred 'Spam' Spafford, the bomb-aimer, as 'a grand guy and many were the parties we had together'. His rear-gunner, Richard Trevor-Roper, silent in the remoteness of the tail, was 'Eton, Oxford', to which his pilot added that 'Trev' 'was probably thinking what I was thinking. Was this the last time we would see England?'

Gibson wrote of his crew as if he knew them intimately, yet in truth this was the first operation that any save wireless-operator Bob Hutchison had flown with him, and it would also be the last. Most of what he stated about the others was wrong. Trevor-Roper attended Wellington College, not Eton, and never Oxford; Pulford, dismissed in the author's original as 'a bit of a dummy', was a Yorkshireman, not a Londoner. Like all 617's engineers, he was a former 'erk', a ground crewman, maid of all work: monitoring the throttles and dials, moving around the aircraft to deal with small problems, check on the rear-gunner or investigate an intercom failure. Every twenty minutes it was his job to log engine temperatures, fuel state. That morning, Sgt. Pulford had received extraordinary permission from Gibson to attend his father's funeral in Hull, an hour's drive from Scampton, to which he had been accompanied by two RAF policemen to ensure that he said not a word to anyone about what he was to do that night.

The pilot described 'Spam' Spafford as a great bomb-aimer, 'but he was not too hot at map-reading'. 'Hutch', the wireless-operator, was 'one of those grand little Englishmen who had the guts of a horse', despite being often airsick. George Deering, the Canadian front-gunner who was a veteran of thirty-five operations, was 'pretty dumb, and not too good at his guns, and it was taking a bit of a risk taking him, but one of our crack gunners had suddenly gone ill and there was nobody else'. If pilot and bomb-aimer had ever caroused together, there is no record of it. For all the 'wingco's' leadership skills, to most of his comrades, and especially to subordinates, little Gibson – on the ground, it was impossible to fail to notice his lack of inches – was a remote figure, respected but not much loved, especially by humbler ranks. A gunner said sourly, 'He was the sort of little bugger who was always jumping out from behind a hut and telling you that your buttons were undone.' By the time Gibson wrote his book, however, both he and Chastise had become legends. Thus, he described a close relationship with his crew as a fitting element of the story.

Reality was that five of the six young men sharing G-George with their squadron commander that night were bleakly aware that they were committed to one of the most hazardous missions of the war, in the hands of a pilot with whom they had never flown over enemy territory. More than that, he was an authentic hero; and heroes are immensely dangerous to their comrades.

Now they were over the North Sea: 'Our noses were going straight for the point at which we had to cross the Dutch coast. The sea was as flat as a mill-pond, there was hardly a ripple … We dropped lower and lower down to about fifty feet so as to avoid radio detection … After a time I tried to light a cigarette. In doing so we again nearly hit the drink and the boys must have thought I was mad. In the end I handed the thing to Pulford to light for me.' Gibson was flying in shirtsleeves, wearing a Luftwaffe Mae West, spoils of war that he had picked up in his fighter days. Although they were operating far below the height at which oxygen was necessary, they were still obliged to wear masks, because these contained microphones for the intercom and VHF link between aircraft – Gibson hankered in vain for throat mikes such as the USAAF employed.

He wrote in 1944, looking back to that unforgettable night: 'One hour to go, one hour left before Germany, one hour of peace before flak. I thought to myself, here are 133 boys who have got an hour to live before going through hell. Some of them won't get back … Who is it will be unlucky? … What is the rear-gunner in Melvin Young's ship thinking, because he won't be coming back? What's the bomb-aimer in Henry Maudslay's ship thinking, because he won't be coming back? … One hour to go, one hour to think of these things, one hour to fly on a straight course and then it will be weaving and sinking to escape the light flak and the fury of the enemy defence.' A few months later, he chose as one of his favourite records on BBC radio's *Desert Island Discs* Wagner's 'Ride of the Valkyries': 'it's exciting, it's grandiose, it's … rather terrible. It reminds me of a bombing raid.' Then Guy Gibson thought about his dog, which was newly dead; and about the epic experience ahead, which would make him one of the most famous fliers in history. 'This was the big thing,' he wrote. 'This was it.'

1

Grand Strategy, Great Dams

1 THE BIG PICTURE

In May 1943 the Second World War was in its forty-fifth month. While it was evident that the Allies were destined to achieve victory over Germany, it was also embarrassingly obvious to the British people, albeit perhaps less so to Americans, that the Red Army would be the principal instrument in achieving this. The battle for Stalingrad had been the dominant event of the previous winter, culminating in the surrender of the remnants of Paulus's Sixth Army on 31 January. The Russians had killed 150,000 Germans and taken 110,000 prisoners, in comparison with a mere nine thousand Axis dead, and thirty thousand mostly Italian prisoners taken, in Montgomery's November victory at El Alamein.

Day after day through the months that followed, newspapers headlined Soviet advances. To be sure, British and American forces also made headway in North Africa, but their drives from east and west to converge in Tunisia embraced barely thirty divisions between the two sides, whereas in the summer of 1943 two million men of Hitler's and Stalin's armies would clash at Kursk and Orel. Axis surrender in North Africa came only on 13 May, months later than Allied commanders had expected.

Almost four years after Britain chose to go to war, and eighteen months after America found itself obliged to do so, the bulk of their respective armies continued to train at home, preparing for an invasion of the continent for which no date had been set. The Royal Navy and the Royal Canadian Navy, assisted by the Ultra codebreakers and latterly by the US Navy, had performed prodigies to achieve dominance over Dönitz's U-Boats: the Atlantic sea link was now relatively secure, and a vast tonnage of new shipping was pouring forth from American shipyards. But this was a defensive victory, its importance more apparent to Allied warlords than to Churchill's and Roosevelt's peoples.

Among the latter, even after its North African successes the standing of the British Army remained low: memories lingered, of so many 1940–42 defeats in Europe, North Africa and the Far East. Many Americans viewed their Anglo-Saxon ally with a disdain not far off contempt. A July 1942 Office of War Information survey invited people to say which nation they thought was trying hardest to win the war. A loyal 37 per cent answered, the US; 30 per cent named Russia; 14 per cent China; 13 per cent offered no opinion. Just 6 per cent identified the British as the hardest triers. 'All the old animosities against the British have been revived,' wrote an OWI analyst. 'She didn't pay her war debts for the past war. She refuses to grant India the very freedom she claims to be fighting for. She is holding a vast army in England to protect the homeland, while her outposts are lost to the enemy ... Phrases such as "The British always expect someone to pull their chestnuts out of the fire" and "England will fight to the last Frenchman" have attained considerable currency.'

Thoughtful British people saw that in almost three years since Dunkirk in June 1940 their army had accomplished relatively little, while the Russians endured the most terrible and costly experiences in the history of war. In 1942 Winston Churchill's reputation as a war leader fell to its lowest ebb, in the face of new British humiliations. The sister of RAF pilot John Hopgood, a future dambuster, was an ATS officer who wrote to her mother in August that she had been cheered by the proclaimed success of the recent Dieppe raid: 'the feeling that at last we were taking the offensive and that my uniform would mean something. A feeling I had very strongly at the beginning of the war, but it has gradually diminished since we have been on the defensive. This is not the talk of a defeatist, but I think it is the truth which most of us have experienced lately. I hope that [Dieppe] was a dress rehearsal and that the performance [D-Day in France] will follow shortly.' Which, of course, it did not.

Belated successes in North Africa delivered Churchill from a real threat of eviction from his post as minister of defence. It was plain that the German flood tide on the Eastern Front was ebbing; that the Russians had survived the decisive crisis of their struggle. But the British people had seen little about their own war effort in which to take pride since the RAF's sublime triumph in the 1940 Battle of Britain. An unusually reflective young airman afterwards looked back upon an early-1943 conversation with friends: 'It was pleasant to sit there and rest a while and think that the worst was behind ... The evening had been pleasant and we had practically "won" the war. But we wouldn't have been so pleased if we had known of the big battles that were to be fought; the heavy casualties to be borne ... The tide had turned, but it was a leap year tide.' This was Guy Gibson, who in those days led 106 Squadron of the RAF's Bomber Command.

It is hard to overstate the impatience felt by millions on both sides of the Atlantic for action against Hitler on a scale to match the efforts and sacrifices of 'Uncle Joe' Stalin's Red Army. The Western Allied leadership was prosecuting the war at a pace that suited themselves, their countries' immunity from invasion from 1941 onwards conferring the luxury of choice, such as the Russians never had, about where and when to engage the Wehrmacht. British caution exasperated US chief of the army George Marshall and his peers. Yet the old prime minister and his chiefs of staff recognised fundamental truths: that Britain and America were sea powers, confronting a great land power. It would be madness to attempt an amphibious return to the continent without command of the air, which could not be secured any time soon. The voice of Churchill, backed by the intellects of his professional military advisers, Portal and Alanbrooke foremost among them, was decisive in delaying D-Day until June 1944, saving hundreds of thousands of American and British lives. Meanwhile, it was vital to the prestige and morale of the Western Allies, and especially those of Churchill's nation, that Britain should be seen to be carrying the war to Germany by any and every means within its power.

2 HARRIS

A consequence of the Western Allies' cautious grand strategy, rendered necessary by their slow industrial build-up, was that the Anglo-American air forces, and especially heavy bombers, constituted their most conspicuous military contribution to the defeat of Germany between the fall of France in June 1940 and the invasion of Normandy. Bomber Command's pre-war estate of twenty-seven British airfields had by 1943 expanded to over a hundred stations, while the RAF's overall strength grew from 175,692 personnel to over a million men, including a significant proportion of the nation's best-educated adolescents.

Some senior officers, including the USAAF's Gen. Carl 'Tooey' Spaatz and Air Chief Marshal Sir Arthur Harris, C-in-C of Bomber Command, believed that air attack on Germany could render redundant a land invasion of the continent. 1943 was the year in which British warmakers drafted plans for Operation Rankin, whereby troops would stage an unopposed deployment to occupy Germany in the event that some combination of bombing, Russian victories and an internal political upheaval precipitated the collapse of the Nazi regime, an abrupt enemy surrender.

Yet, while Winston Churchill committed a lion's portion of Britain's industrial effort to the air offensive, he never shared the airmen's extravagant hopes for it. For a season in 1940, when Britain's circumstances were desperate, he professed to do so. Once the threat of invasion receded, however, he recognised that, while bombing could importantly weaken the German war machine, it could not hope to avert the necessity for a continental land campaign. The airmen's most critical contribution until June 1944 was to show Churchill's people, together with the Americans and – more important still – the embattled Russians, that Britain was carrying the war to the enemy. The prime minister recognised, as his chiefs of staff often did not, the value of 'military theatre' – conspicuous displays of activity that sustained an appearance of momentum, even when real attainments were modest. As the author has written elsewhere: 'There must be action, even if not always useful; there must be successes, even if overstated or even imagined; there must be glory, even if undeserved.' Through those apparently interminable years between Dunkirk and D-Day, again and again the BBC prominently featured in its news bulletins the words 'Last night aircraft of Bomber Command ...' followed by a roll call of industrial targets attacked in France, Italy, and above all Germany.

In 1940–41 the RAF caused mild embarrassment to the Nazi leadership, which had promised to secure the Reich against such intrusions. Bombing nonetheless inflicted negligible damage upon Hitler's war effort. Although more aircraft became available during the winter of 1941, poor weather, navigational difficulties and German fighters inflicted punitive casualties upon the attackers, who still made little impact on the enemy below. Thereafter, however, a succession of events took place which progressively transformed the offensive.

In December 1941 the prime minister and the Air Ministry received an independent report from the Cabinet Office, commissioned by Churchill's personal scientific adviser Lord Cherwell, the former Professor Frederick Lindemann, analysing the effectiveness of British bombing through a study of aiming-point photographs returned by aircrew. This devastating document showed that the average RAF crew on an average night was incapable of identifying any target smaller than a city. In consequence, and after a vexed debate in which practical issues dominated and moral ones did not feature at all, British strategy changed. By a decision for which Cherwell was prime mover in concord with Air Chief Marshal Sir Charles Portal, since October 1940 head of the air force, it was agreed that instead of pursuing largely vain efforts to locate power stations, factories and military installations, the RAF would assault entire urban regions.

The principal objective would be to 'de-house' and frankly terrorise the German industrial workforce – break the spirit of Hitler's people – even though the Luftwaffe had conspicuously failed to achieve this against Churchill's nation. The new policy, known as 'area bombing', was never directly avowed to the public, nor indeed to Bomber Command aircrew, who were told that the RAF continued to strike at military and industrial targets, with civilian casualties an incidental, and implicitly regrettable, by-product. This was a falsehood. Between 1942 and 1945, the civilian population of Hitler's cities was the target of most British bombing.

America's entry into the war in December 1941 made eventual Allied victory seem certain. Until a continental land campaign began, US air chiefs were as eager as their British counterparts to demonstrate their service's war-winning capabilities. Daylight operations by American B-17 Flying Fortresses and B-24 Liberators began slowly to reinforce the RAF's night campaign. The British received early deliveries of a new generation of four-engined heavy bombers – Short Stirlings and Handley-Page Halifaxes, followed by Avro Lancasters – which progressively increased Bomber Command's striking power. They also acquired 'Gee', the first of a succession of electronic aids which improved the accuracy of RAF navigation.

Finally, in February 1942 Sir Arthur Harris became commander-in-chief of Bomber Command. Britain's interservice wrangles and clashes of personality inflicted less damage than did those of the United States, and for that matter Germany, upon their own war efforts. They nonetheless absorbed time and energy. The Royal Navy and the RAF disliked and distrusted each other as a matter of course, rivals for resources in an ongoing struggle in which both were frequently rebuked by the prime minister. Many airmen also viewed soldiers with the disdain due to their serial record of defeats.

Harris became the most intemperate squabbler. He regarded with contempt Special Operations Executive, the covert warfare organisation for which a handful of bombers was grudgingly committed to drop arms to the Resistance in Occupied Europe. As for sailors, it was one of his favourite sayings that the three things one should never take on a boat were an umbrella, a wheelbarrow and a naval officer. He fought like a tiger against the diversion of heavy aircraft to support the Battle of the Atlantic, arguing that it was a far more economical use of force to bomb U-Boats in their German construction yards than to waste flying hours searching for them in the vast reaches of the oceans. He described the RAF's Coastal Command as 'an obstacle to victory', despite the importance of its Very Long Range Liberator squadrons in countering U-Boats. Meanwhile the admirals, who had a good case in pleading for more aircraft, spoilt this by insisting that Bomber Command should repeatedly attack the Germans' concrete submarine pens on the north-west French coast, which were invulnerable to conventional bombs, and heavily defended by flak and fighters.

Harris waged a further ongoing struggle with the Air Ministry, of which much would be seen in the debate about Germany's dams. The C-in-C of Bomber Command was an elemental force, single-minded in his conviction that he, and he alone, could contrive the defeat of Nazism through the systematic, progressive destruction of Germany's cities. Alan Brooke, chief of the British Army, recorded characteristic Harris testimony at a chiefs of staff meeting: 'According to him the only reason why the Russian army has succeeded in advancing is due to the results of the bomber offensive! According to him ... we are all preventing him from winning the war. If Bomber Command was left to itself it would make much shorter work of it all!'

Freeman Dyson, a brilliant young scientist who spent much of the war in the Operational Research section of Bomber

Command at High Wycombe, characterised his chief as a 'typical example of a prescientific military man ... brutal and unimaginative'. Hyperbole was this glowering figure's first choice of weapon in exchanges with those who crossed him. This became a kind of madness, and Harris a kind of madman, but in the unwelcome predicament of Britain for much of the Second World War, Churchill recognised that such a figure had important uses. Horace Walpole wrote in the mid-eighteenth century: 'No great country was ever saved by good men, because good men will not go to the lengths that may be necessary.'

Though Harris became the foremost exponent of 'area bombing', which has ever since been inseparably identified with his name, he was not its begetter, merely its obsessive implementer. It was widely believed, especially by soldiers and sailors, that Bomber Command's C-in-C achieved an intimacy with Churchill, by exploiting the proximity of Chequers to his headquarters at High Wycombe, to secure support for his purposes. This view seems unfounded. The prime minister after the war described the airman as 'a considerable commander'. He rightly judged that Harris instilled in the bomber offensive a dynamic, a sense of purpose, which it had previously lacked. He valued the airman's skilful exploitation of public relations, conspicuously manifested in his May–June 1942 'Thousand Bomber raids', of which the most famous, or notorious, was directed against Cologne.

Yet the prime minister never much liked 'Bert' Harris – as he was known to intimates. 'There was a certain coarseness about him,' Churchill observed, implicitly contrasting the airman, who set no store by social graces, with such officers as Sir Harold Alexander, a gentleman in every respect, who became Churchill's favourite general. Harris, just short of fifty when he assumed command, was the son of an engineer in the Indian Civil Service. He spent much of his youth in southern Africa, and especially Rhodesia, which he came to love. The reverse of the coin of his force of character was a vulgarity of language and behaviour, exemplified by his observation that Britain's generals would take tanks seriously only 'when they learned to eat hay and fart'.

He experienced a lunatic moment in January 1943, when he became so incensed by the incidence of venereal disease among aircrew that he issued an edict, without consultation, that every diagnosed sufferer should be obliged to restart from scratch his tour of thirty 'trips' to Germany. This monstrous threat, rooted in a notion that shirkers were inviting infection in order to escape from operations, was withdrawn only in June, following the intervention of Sir Archibald Sinclair, Secretary of State for Air, who overruled the C-in-C.

Nonetheless, at a time when many others to whom Churchill entrusted high commands – for instance Dill, Wavell, Auchinleck – had proved weak vessels, despite their impeccable manners, Harris, a four-letter man in the eyes of most of his peers, possessed qualities that the prime minister valued. He said long afterwards of Bomber Command's chieftain, in conversation with his last private secretary: 'I admired his determination and his technical ability. He was very determined and very persuasive on his own theme. And the Prof. [Lord Cherwell] backed him up. You must remember that for a long time we had no other means than Bomber Command of hitting back. The public demanded action and rejoiced at our counter-blows at German cities after Coventry and so many other towns … Large numbers of German aircraft and vast resources of manpower and material were tied up in their air defence.'

Harris's personal life was unorthodox. His recreation was driving ponies: he had been known to take the reins of his own trap to travel to Chequers to see the prime minister; if called upon, he could manage a four-horse team. His wife Barbara walked out on him in 1934, securing a contested divorce on grounds of his adultery. Thereafter, their three children as well as herself were ruthlessly written out of the script of his life, and even out of his subsequent official biography. Not long afterwards he proposed marriage, down the intercom of a Hornet Moth biplane in Rhodesia, to a very young woman to whom he was giving a joyride. 'I think it would be very nice if you were to marry me – will you?' Harris demanded of Polly Brooks, who – in P.G. Wodehouse's phrase – turned him down like a blanket. Miss Brooks offered the reasonable excuse that the airman was old enough to be her father, though she added politely, 'It's very nice of you to ask me.'

Instead, in 1938 he married another very young woman, twenty-three-year-old Therese Hearne, a strong-minded Catholic always known as Jill, who gave birth to a daughter, Jackie, the following year. Thus, through the years during which Harris directed Britain's bomber offensive from his High Wycombe headquarters, at his official residence in nearby Springfield House a wife more than twenty years his junior entertained a procession of Allied warlords while rearing a small child.

Conflict was Harris's environment of choice, his feuds tempered only by a harsh wit. He once scrawled on a memorandum describing complex alternative means of destroying a target: 'TRY FERRETS'. He enjoyed the joke against himself of being stopped for speeding in his Bentley – an offence which he revelled in repeating – and rebuked by a policeman who told him that he might have killed someone. 'Young man,' the air marshal replied, albeit surely apocryphally, 'I kill thousands of people every night!' His staff and close associates were unable to decide whether the chronic ulcers from which he suffered stimulated his ill-temper, or were precipitated by it.

During the year since the new C-in-C assumed direction of Britain's strategic air offensive, he had transformed Bomber Command from a transport service dumping ordnance almost indiscriminately around the German countryside into a serious weapon of war. Sceptics, some of them within the RAF, sustained doubts about whether burning cities was doing anything like as much as Harris claimed to advance Allied victory. Sir Wilfred Freeman, Portal's able vice-chief, wrote to the CAS on 16 September 1942 deploring the grossly exaggerated claims made by some commanders: 'in their efforts to attract the limelight, they sometimes exaggerate and even falsify facts. The worst offender is C-in-C Bomber Command.'

Nonetheless, the RAF's publicity machine made much of 'Bomber' Harris, as he was nicknamed by the press, and of the devastation that his aircraft inflicted nightly upon Germany. In 1940 Bomber Command dropped just 13,033 tons of bombs on enemy territory; in 1941, 31,704 tons. Thereafter, under Harris's command, in 1942, 45,561 tons fell; in 1943, 157,457 tons; in 1944, 525,718 tons. By the war's end, Bomber Command was capable of raining upon Hitler's people in a single twenty-four-hour period as many bombs as the Luftwaffe dropped during the course of its entire 1940–41 blitz on Britain.

Autocratic is an inadequate word to describe Harris's style of command. He considered himself to have been entrusted with a vast responsibility, and resisted any interference, criticism or even interrogation about his manner of fulfilling this. He regarded with contempt the Directorate of Bomber Operations, a cell within the Air Ministry which provided

Portal with in-house advice that often ran counter to the convictions of Harris and his staff, few of whom dared to think for themselves, far less speak out. He especially loathed Gp. Capt. Syd Bufton, who had successfully championed the 1942 creation of an elite Pathfinder force – what became Bomber Command's No. 8 Group – against the opposition of the C-in-C. 'Morning, Bufton,' he once greeted that officer on arriving at the Air Ministry for a meeting. 'And what have you done to impede the war effort today?'

Among the terms of abuse Harris heaped upon his critics, that of 'panacea merchant' was intended to be the rudest, embracing Bufton, sometimes Portal, even the prime minister. The words meant that a given individual was advocating means of defeating the Axis, or more especially of bombing Germany, which did not require the systematic demolition of its urban centres. The relationship between Harris and Portal was extraordinary. Bomber Command's C-in-C frequently defied direct instructions from the Air Ministry, and sometimes from Portal himself, to divert aircraft from attacking cities towards alternative objectives, of which dams came to be among the most contemptuously regarded, alongside ball-bearing factories, V-weapon sites, French railways, synthetic-oil plants, aircraft factories and U-Boat pens.

The head of the RAF was subjected to barrages of invective from his nominal subordinate, to which he was often driven to respond in the language of a headmaster rebuking an errant pupil. In April 1943 there was a characteristic Harris explosion, about a pamphlet circulating widely in British cities and allegedly also at some bomber stations, headed 'STOP BOMBING CIVILIANS', together with a demand from the C-in-C for the identification and indictment for treason of its authors, essentially for highlighting inconvenient truths.

Portal replied on the 9th: 'It does not appear that prosecution of the authors for circulating [this pamphlet] among civilians would have the slightest chance of success. No court would be likely to hold that it was an offence to advocate that bombing should be confined as far as possible to military objectives. You suggest that this pamphlet comes under the heading of subversion when addressed to an individual in the Service. Even if this is technically correct I do not think it would be prudent to maintain in public that a pamphlet such as this, maintaining a moderately-worded statement of the case against civilian bombing, is likely to incite aircrew to disobey orders … We can however reduce the likelihood of such opinions gaining ground by emphasizing in our publicity industrial damage rather than the destruction of civilian dwellings.' The RAF's chief of staff replied to another incontinent note from the C-in-C of Bomber Command: 'I feel bound to tell you frankly that I do not regard it as either a credit to your intelligence or a contribution to winning the war. It is in my opinion wrong in both tone and substance.'

How did Harris retain his job until 1945, when he displayed an unreason and insubordination that few other senior officers would have dared to indulge? He possessed in full measure the quality of 'grip' indispensable to successful commanders in war. Propaganda elevated Harris into a famous figure, and such people become ever harder to sack. He was a man of steel, certain of his purposes when many others, including Portal, wavered and doubted about how the air offensive should best be conducted. 'Peter' Portal, as he was known to intimates, possessed an intellect unusual among service officers of any rank, including chiefs of staff. He was a brilliant diplomat, especially in conducting relations with the Americans, whom Harris privately regarded with contempt. But he was also often indecisive. Portal did not oppose area bombing, indeed presided over its inception. He merely favoured leavening fire-raising attacks on cities with precision strikes whenever suitable targets could be identified, and means found to hit them.

Nobody in high places was sufficiently assured of the superior merit of any alternative strategy, or of any more effective commander at High Wycombe, to remove Harris. Later in the war, extraordinary though it may seem when hundreds of bombers continued to fly forth nightly to broadcast death and destruction, the prime minister lost interest in the air offensive: it is striking how little mention Bomber Command receives in the final volumes of Churchill's memoirs. Once the great land campaigns got under way, armies and the fate of nations entirely eclipsed air forces as the focus of his attention. In the early months of 1943, however, Harris was near the zenith of his fame and importance. He was playing a role more conspicuous than that of any other British commander towards encompassing the destruction of Nazism. Without Harris, without Bomber Command, until June 1944 there would have been only Gen. Sir Bernard Montgomery and his Eighth Army, the North African and thereafter Italian 'sideshows'.

In January 1943, when President Franklin Roosevelt and the US chiefs of staff met Churchill and the British chiefs in newly liberated Casablanca, the British achieved one of their last diplomatic triumphs of the war, before American dominance of policy and strategy became explicit. The US team unwillingly accepted that there would be no Western Allied invasion of north-west Europe that year. Instead, there would be amphibious assaults on Sicily and probably thereafter Italy, together with a 'combined bomber offensive' on Germany by the two air forces. The consequent so-called Casablanca Directive ordered British and American air chiefs: 'Your primary aim will be the progressive destruction and disloca- tion of the German military, industrial and economic system, and the undermining of the morale of the German people to a point where their capacity for armed resistance is fatally weakened.'

In the event, no 'combined' offensive took place; instead, there was a competition between the US and British air forces. Sir Arthur Harris paid mere lip service to Casablanca's empha- sis on refined targeting. His aircraft continued to heap fire and destruction on Germany's cities by night, while in daylight the USAAF claimed to pursue precision bombing of identified weak points in the Nazi war economy. Because, in reality, American bombing proved highly imprecise, especially in poor weather, hapless German civilians saw little distinction between the rival strategies. Moreover, intelligence about enemy industry remained a weakness of the strategic air offen- sive from beginning to end.

In February 1943, Harris stood on the brink of a new campaign to deploy almost the entire resources of his Command against the industrial cities of north-west Germany, which would become known as the Battle of the Ruhr. The enthusiasm of others – within the Ministry of Economic Warfare, the Air Ministry, the Admiralty and even Downing Street – for more selective targeting roused his scorn. Portal, Bufton and some other airmen regarded extremely seriously the Casablanca Directive: they strove to identify and attack choke points in the German industrial machine. Any such proposals, however, encountered savage resistance from Harris's headquarters at High Wycombe to the diversion of aircraft from 'area bombing'. Until the last weeks before Guy

Gibson and his men set forth on what would become the most applauded operation of Bomber Command's five-year campaign, its commander-in-chief wanted 'his' aircraft, 'his' offensive, to have nothing to do with it.

3 THE 'PANACEA MERCHANTS'

As far back as October 1937, RAF planners identified Germany's water resources – dams and reservoirs – as a vulnerability in the Nazi industrial machine. Bomber Command initially focused on nineteen Ruhr power stations and twenty-six coking plants, which its staff believed could be destroyed by three thousand bombing sorties. This would allegedly bring Nazi war production to a standstill, in return for an anticipated loss of 176 British aircraft. Then the Air Targets Sub-Committee of Major Desmond Morton's Industrial Intelligence Centre took a hand, highlighting the dependence of electricity generation, mining and coking activities upon water supply. Morton's cell urged that the mere breaching of two dams, the Möhne and the nearby Sorpe, could achieve the desired outcome, a prospective war-winner, for far less expenditure of effort.

It was meaningless, however, for planners to focus upon any target system unless means existed to attack it, which they certainly then did not. Destruction of the Möhne, composed of almost a million cubic feet of masonry, was admitted to be 'highly problematic' at a time when the RAF's heaviest aircraft, the Heyford, carried only 500-lb bombs. In March 1938 it was agreed that the significance of the dams made it worth exploring 'newly-developed weapons' with which they might be attacked. But, as the engineer Barnes Wallis often later complained, while the RAF was doctrinally committed to strategic bombing, its planners gave negligible attention to the ballistic challenges involved in the demolition of large structures.

Britain's airmen were imbued with a faith that the mere fact of subjecting an enemy nation to bombing would cause its people, and even perhaps its industrial plant, to crumble before them. 'In air operations against production,' wrote Gp. Capt. John Slessor, a future head of the RAF, in 1936, 'the weight of attack will invariably fall upon a vitally important, and not by nature very amenable, section of the community – the industrial workers, whose morale and sticking power cannot be expected to equal that of the disciplined soldier.' The intention of Britain's airmen to play a decisive and independent role in achieving victory over Germany relied for fulfilment more on expectations of the havoc that terrorisation of civilians would wreak than on any rational analysis of the weight of attack necessary to cripple the Nazi industrial machine.

In the cases of the Möhne and the Sorpe, experts at the Air Ministry's research department at Woolwich warned: 'If the policy to attack dams is accepted, the [Ordnance] Committee are of the opinion that the development of a propelled piercing bomb of high capacity would be essential to ensure the requisite velocity and flight ... Even then its success would be highly problematical.' Subsequent debate concluded that, given the low standard of aiming accuracy achieved by RAF bomber pilots in peacetime, before they were even exposed to enemy fire, a successful attack on Germany's dams was impracticable.

Yet British planners continued to nurse this dream, which was mirrored by a German nightmare. On the eve of war, 29 August 1939, Justus Dillgardt, chairman of the *Ruhrtalsperenverein* organisation responsible for regional water resources, suggested that a succession of bombs exploded sub-aquatically within a hundred feet of the base of his dams, the effects multiplied by the water mass – a scientific phenomenon identified by the British only three years later – might prompt dam collapses. If the Möhne was breached, Dillgardt warned that 'this entire industrial area ... would be completely paralysed ... not only would the population of four to five millions be without water, but all mines and coking plants would suddenly cease work owing to lack of industrial water supply'.

Through the early war years, both the Air Staff and British economic warfare researchers sustained enthusiasm for striking at the enemy's water resources. The Ruhr and its industries accounted for a quarter of the Reich's entire consumption, much of this derived from the Möhne reservoir. The Air Targets Sub-Committee was told that its destruction would create 'enormous damage', affecting hydro-generating stations. The 'low-lying Ruhr valley would be flooded, so that railways, important bridges, pumping stations and industrial chemical plants would be rendered inoperative'.

In June 1940, amid the crisis of the fall of France, Bomber Command groped for means of striking back at the victorious Nazis. Its senior air staff officer, Gp. Capt. Norman Bottomley, again urged the merit of breaching enemy dams. On 3 July, Portal, then Bomber Command's C-in-C, wrote to the Air Ministry pinpointing a key target. Surely, he said, 'the time has arrived when we should make arrangements for the destruction of the Möhne Dam ... I am given to understand that almost all the industrial activity of the Ruhr depends upon [it].' He suggested a possible torpedo attack by Hampden bombers, or high-level bombing attack against the 'dry' side of the Möhne's wall.

But what realistic prospect could there be of such an operation's success? Throughout the wars of the twentieth century, again and again it was shown that strategy must follow technology, and also required mass, whether in the matter of landing-craft for amphibious operations; radar to negate prime minister Stanley Baldwin's warning – correct when it was delivered in 1932 – that 'the bomber will always get through'; or attacks on German dams. It was useless to identify a purpose unless means to fulfil it existed, or might be created, as they certainly could not have been in 1940, when Bomber Command was weak, and Britain's war effort was focused upon averting defeat.

The weapons available to pursue the airmen's ambitions remained inadequate in quality as well as quantity. The Hampden bombers mentioned by Portal could carry nothing like heavy enough ordnance to dent, far less breach, the Möhne. A December 1940 study of German bombing of the UK concluded, a trifle ungrammatically: 'comparison of the Results … with that obtained by [RAF] General Purpose bombs against similar enemy targets left no doubt as to the inefficiency of our bombs'. GP bombs contained too little explosive, yet until the end of the war Britain's airmen continued to risk and sacrifice their lives to drop more than half a million of them. The 4,000-lb HC – High-Capacity – bombs introduced in 1941, and known to aircrew as 'cookies', were relatively efficient wreckers of urban areas: sixty-eight thousand were dropped in the course of the offensive. They were useless, however, against such huge structures as dams.

It was noted that during the Spanish Civil War, Nationalist forces with direct access to the Ordunte and Burguillo dams had failed in attempts to destroy them, using charges laid by hand. Winston Churchill urged the American people: 'Give us the tools and we will finish the job.' Where, now, were to be found tools such as might collapse the enemy's dams? The

Deputy Chief of Air Staff responded to Portal's proposal to target the Möhne that to breach its concrete mass would require dropping a barrage of a hundred mines on the reservoir side, together with a huge bomb exploded on the 'dry' side. 'The practical difficulties of this method are considered to be insuperable at present.'

The great dams of north-west Germany possess a beauty unusual among industrial artefacts, a majesty enhanced by their settings among hills and woodlands. They were created in the early twentieth century, to meet soaring demand for water from both a rising population and burgeoning regional heavy industries. Work on building the Möhne, then the largest structure of its kind in Europe, began in 1908, part of a construction marathon undertaken by the *Kaiserreich* in the decade before the First World War. Completed in 1913 at a cost of 23.5 million marks, it was opened by Wilhelm II. The flooding of its reservoir, which eventually held almost five thousand million cubic feet of water, interrupted the Möhne river, which flowed west into the Ruhr.

The new lake, the Möhnesee, displaced seven hundred rural inhabitants of the Sauerland. It became a focus of national pride, and also a tourist attraction. Floatplanes and pleasure craft plied its waters between the wars, as they did likewise on its near neighbour the Sorpetalsperre, constructed between 1926 and 1935 in the Sorpe river valley below the village of Langscheid. A narrow-gauge railway carried more than 300,000 tons of material to what became Europe's largest construction site, where an earthen embankment was raised around a masonry core. On completion, this rose 226 feet above the river, and was 2,297 feet in length. Its lake held 2,520 million cubic feet of water.

The Eder, forty-five miles south-eastwards in Hesse, was a smaller structure, 157 feet high and 1,312 feet wide, damming the Eder river two miles from the towering old castle of Waldeck to create a reservoir seventeen miles long, holding just over seven thousand million cubic feet of water. It was another pre-World War I project, opened in 1914, and quite unrelated to the Ruhr system. The Nazis stocked all their big reservoirs with quantities of fish, especially pike and carp, as a strategic resource in anticipation of falling sea catches when hostilities began.

The German dams, and especially the Möhne, retained their fascination for British air and economic intelligence officers through the early war years. Many and many were the hours that experts pored over photographs and technical reports about wall thicknesses, surrounding topography, defences. W/Cdr. C.R. Finch-Noyes of the Government Research Department at Shrewsbury examined all available studies and assessments, and on 2 September 1940 reported that if ten tons of explosive could be detonated beneath the Möhne's wall, 'there seems a probability that the dam would go'. He suggested that if aircraft flying at 80 mph and very low altitude released a succession of one-ton units close to dams, natural momentum would propel them across the water to the walls, where they would sink, to be detonated by hydrostatic pistols. This was a fascinating notion, because it anticipated – quite unknowingly – Barnes Wallis's 'bouncing bomb', in truth a depth-charge. Finch-Noyes indeed urged that a standard British naval depth-charge might fulfil the requirement. One of his old colleagues, a strange, erratic air pioneer named Noel Pemberton-Billing, suggested that the dam might be attacked using a 'hydroplane-skimmer' that would jump over its buoyed net defences.

The Admiralty showed an interest in these proposals, not to attack dams but instead to develop weapons that might destroy targets of its own, enemy warships prominent among them. Some serious research and tests were conducted. On 2 April 1941, a paper authored by Finch-Noyes – 'Memorandum of proposed Methods of Attacks of Special Enemy Targets' – highlighting dams, was circulated among service departments. After studying this document, however, Bomber Command's senior operational staff officer wrote that Finch-Noyes's ideas must founder because of the immense tonnage of explosive needed to breach a big dam, such as no existing or prospective British aircraft could carry. The project might be of more value to the Royal Navy, he said, because smaller versions of Finch-Noyes's proposed weapon could be practicable for use against warships.

It is often the case with big ideas, especially scientific ones, that several individuals or institutions grasp the same one independently, sometimes continents apart. Both sides in the most terrible war in history recognised reservoirs as significant industrial targets. In 1940 the Luftwaffe considered attacking the Derwent and Howden dams near Sheffield. The German airmen eventually abandoned consideration of such a strike, for the familiar reason that the dams seemed too large to be breached with existing weapons. It required the advent of a white-haired fifty-five-year-old visionary to empower the Royal Air Force to address a challenge that had vexed and frustrated its leaders since 1938.

2

The Boffin and His Bombs

1 WALLIS

If Germany's dams had been attacked with conventional bombs, rockets or shells, posterity – at least British posterity – might have taken little heed of the story. As it was, the means employed, and the man who devised them, confer enduring fascination. 'Among special weapons,' recorded a post-war study of RAF armament by the service's Air Historical Branch, in language that reflects self-congratulation, 'the "Dam Buster" must take pride of place … the story of its development and production is an epic in the history of aerial bombs.'

Barnes Wallis was the only 'boffin' – to be more accurate, he was an engineer – to achieve membership of Britain's historic pantheon of World War II, behind Winston Churchill but alongside the Ultra codebreaker Alan Turing and the fighting heroes of the conflict. Until 1951, when Paul Brickhill's book was published, scarcely anyone knew or remembered anything of Wallis. He had enjoyed some celebrity in the pre-war years, especially in connection with his work on the great airship R100. From 1939 onwards, however, he vanished behind a curtain of official security. He became famous only in the decades that followed, after the release of the film *The Dam Busters*, in which he was portrayed by Michael Redgrave.

The Wallis legend depicts a genius, seized with a potentially war-winning idea, fighting a lone battle against unimaginative bureaucrats to achieve fulfilment of his conception. The truth was almost entirely the other way around. What was extraordinary about the concept of what became known as the 'bouncing bomb' was that in the midst of an existential struggle in which Britain was striving with meagre resources, suffering repeated defeats and setbacks, some of the guiding lights of the war effort, both servicemen and civilians, grasped the potential of Wallis's fantastic idea, supported its evolution, and within a few short weeks of securing command approval contrived the manufacture of workable examples. Moreover, officialdom proved astonishingly – indeed naïvely – willing to share the inventor's extravagant hopes for the impact of such an assault upon the Nazi war machine. While scepticism had to be overcome about whether Wallis's weapons would work and whether resources could be found to construct them, there was much less rigorous analysis of how drastically breaking dams would harm the interests of Hitler – except by Sir Arthur Harris, who had locked himself into a narrative of his own.

Even had Barnes Wallis failed to secure fame through breaching the dams, he would have deserved notice as a remarkable human being. He was the son of a doctor who practised in London's humble New Cross area, hampered by having suffered polio. Barnes, born in 1887, attended a minor public school, Christ's Hospital, only after winning a scholarship. He missed university, and instead became an engineering apprentice at a shipbuilding firm, then in 1913 graduated to working on airship development for the industrial giant Vickers. While he was adequately paid, his finances were chronically strained by his insistence upon aiding other members of his unlucky family. No stranger to pawnbrokers' shops, he once sold a bicycle to pay for his parents to enjoy a holiday.

When World War I came, Wallis's repeated attempts to join the army foundered because Vickers reclaimed his services. He served for a few months on airships, with the rank of sub-lieutenant in the Royal Naval Air Service, but retained a lifelong guilt that he had not fought as did most of his contemporaries. In 1922, with the post-war run-down of the armed forces, Vickers abandoned airship production and made Wallis redundant. During the years that followed, somewhat unexpectedly he served as a part-time Territorial Army soldier in an anti-aircraft artillery unit. For a time he studied for an external degree at London University, and was reduced to seeking employment through the scholastic agency Gabbitas Thring, who found work for him as a mathematics teacher at an English school in Switzerland. It was from there, as a bachelor already thirty-five, that he began writing to his seventeen-year-old cousin by marriage, Molly Bloxam, to whom he explained mathematical formulae, then progressed to discussing technical and physics issues that fascinated him. Their correspondence developed into a romance. In that long-ago era before social telephoning, he wrote ten- to twelve-page letters to his beloved, signing himself 'your affectionate cousin'.

Molly soon pledged her heart to Barnes, but her father was wary of this middle-aged 'cradle-snatcher', for a year restricting their correspondence to a letter apiece a fortnight. It was only in April 1925, three years to the day after they met, that they were finally married – to live happily ever after. The Wallises never became rich, but in 1930 they achieved chintzy middle-class comfort in a mock-Tudor suburban home at

Effingham in Surrey which eventually hosted the rumpus generated by four noisy children. This largely self-taught poly-math rejoined Vickers in the year of his wedding as assistant chief designer, working on the R100, then the largest airship ever built. While labouring at this day job he found time for bell-ringing in the village church and service on the parish council – he was a devout Christian and vegetarian. Until war came he practised Sunday observance, declining even to read a newspaper. A friend wrote of a conversation with Wallis in which he exuberantly expressed his admiration for God: 'My dear boy, do you realise that the Almighty has arranged a system whereby *millions* of electric circuits pass up and down a single cord no bigger than my little finger, and each one most beautifully insulated. The spinal cord is an absolute marvel of electronics!' So deep was Wallis's attachment to family that for most of his life he subsidised, and indeed supported, first his father, then his grown-up sister and her husband.

All the Wallises were music-lovers, and Barnes played an occasional round of golf on the course adjoining his garden. He proved an ingenious handyman around the house, and an imaginative wood-carver. Though not teetotal, the family drank little, and never succumbed to extravagance. Barnes took a cold bath every morning. For all his devotion to Molly, he could be a stern paterfamilias. 'We were used to my father isolating himself in his study at the top of the house,' said his daughter Mary. 'Always working, often abstracted, he was frequently absent from the daily round of chat, laughter and games which large families enjoy. But when he did join in it was lively and great fun. Even in the darkest days he would burst into cheerful, spontaneously made-up doggerel verse under the name "Spokeshave-on-Spur", which delighted us all.' Wallis relaxed discipline on annual family camping and walking holidays. Mary described how, on a Dorset beach, he taught his children to skim flat stones: 'Mine went plop, plop and sank. His would slide smoothly with six or seven hops and quietly submerge.' Barnes and Molly, their daughter added, 'succeeded in protecting us from fear, anxiety, hunger or distress', a notable achievement for any parents.

Yet Wallis's stubborn, spiky eccentricities not infrequently engaged him in quarrels. There was a peculiar episode when Molly met, admired and brought home to Effingham the great birth-control evangelist Marie Stopes. She and Barnes disliked each other on sight, and continued to do so, though her son Harry eventually married the engineer's daughter Mary. At the outset, Wallis and Stopes argued fiercely over his indulgence and indeed encouragement of Molly's semi-overt breastfeed-ing of her baby of the moment, a practice which the visitor deemed barbaric.

Barnes's favourite domestic relaxation was to read aloud to Molly from Dickens, Hardy or Jane Austen while she mended the children's clothes. The Wallises were *good* people, if that is not an inadequate adjective, committed to the virtues of honesty, family and honourable behaviour. This tall, angular figure was also, of course, a workaholic. 'He was a collision of times,' observes Richard Morris. 'In manners and values he was of the 1890s; in aerodynamic possibility, of the 2030s or beyond. He combined confidence, self-pity, vision, regret, hope, loyalty, disdain and ten-score other characteristics.'

To understand Wallis's wartime experiences it is important to recognise that, while his talents and imagination were remarkable, he was very far from right about everything. All his life he pursued doomed projects with the same manic, obsessive commitment that he brought to those that pros-pered. Throughout a long association with airships, he failed to perceive that winged aircraft represented the future, writing to a colleague soon after World War I: 'All my heart is in airships, and I *have* worked so hard.' He championed their cause, and especially that of the R100, even after the 1930 incineration of the R101, together with similar disasters in the United States, had laid bare inherent limitations of the lighter-than-air concept.

In 1933 the M.1/30, a prototype torpedo biplane which Wallis designed, broke up in mid-air, though the structural failure was not his fault. Its test pilot, Captain Joe 'Mutt' Summers, took to his parachute successfully, but the plane's observer had a close brush with death when his straps became entangled with the rear machine-gun as the wreck screamed earthwards. The man was fortunate to escape, and to deploy his canopy, before the plane spun into the ground. While Wallis was often applauded for creating the geodetic framework of the Wellesley and Wellington bombers – latticing derived from his wiring system for harnessing the gasbags of airships, which created exceptional fuselage strength – other nations concluded that it was too complex to be cost-effective, and the RAF spurned geodetic frameworks for its later heavy bombers.

Between 1941 and 1943 the foremost brains of Vickers-Armstrong were engaged in creating a new aircraft, christened the Windsor, armed with 20mm cannon, capable of carrying a bomb load of fifteen tons at a speed of 300 mph. Rex Pierson, Barnes Wallis – who held the title of Assistant Chief Designer (Structures) – and supporting teams of engineers and draughts-men devoted countless hours to this project, which never advanced beyond the prototype stage. The ever-improving performance of the Avro Lancaster, which entered service in 1942, made the Windsor redundant, though work on it contin-ued through 1944.

None of the above is intended to detract from Wallis's achievements – merely to explain why it was not unreasonable for those in authority to greet with caution his higher flights of imagination. At one time and another of his life, large sums of public money were expended on the development of devices, weapons, and indeed aircraft which failed after he had proclaimed their virtues at Whitehall meetings with the same messianic fervour he deployed in advocating his winners.

Moreover, Wallis was only one among a host of enthusiastic inventors peddling ambitious schemes to the armed forces. Lord Cherwell, the prime minister's favourite scientist, railroaded into the experimental stage an absurd scheme for frustrating enemy aircraft with barrages of aerial mines. Cherwell likewise promoted a CS – Capital Ship – bomb that was an expensive failure, as were early British AP – Armour-Piercing – bombs. Lord Louis Mountbatten, as director of combined operations, sponsored a scheme for creating aircraft-carriers contrived from ice blocks. Barnes Wallis attempted to persuade the Royal Navy to adopt a smoke-laying glider of his invention. The Americans conducted experiments in fitting incendiary devices to bats, to be dispatched over enemy territory, an abortive operation codenamed X-Ray. Evelyn Waugh's description, in his satirical war novel *Put Out More Flags*, of Whitehall recruiting a witch doctor to cast spells on Hitler, did not range far beyond reality. Aircraft designer Norman Boorer said: 'There were many, many crazy ideas being put forward by all sorts of scientists.'

Despite Wallis's white hair and the faraway look that was often in his eyes, he was anything but unworldly – indeed, he might be considered a veteran 'Whitehall warrior'. Over two decades of nurturing and supervising complex projects he had honed skills in haranguing committees; guile in exploiting personal relationships; boldness in bullying companies and institutions to assist him in pursuing his purposes. Like many brilliant men, he existed in a default condition of exasperation towards the failure of others to see things as he did. In 1940, when he was working on modifications to the Vickers Wellington and also on a six-engined 'Victory' bomber of his own conception, he wrote to an old World War I colleague: 'Life is almost unrelieved gloom – worse than 25 years ago, except that this time I can feel that I am doing something useful whereas last war I certainly was not ... Tremendously busy – on big developments, which if they had been put in hand two years ago would have won us the war by this time. Too late as usual.'

His 'Victory' bomber, claimed Wallis in July 1940, 'is going to be the instrument which will enable us to bring the war to a quick conclusion'. Since these aircraft would operate at an altitude beyond the reach of German fighters, they could fly 'at their leisure and in daylight ... Irreparable damage could be inflicted on the strategic communications of the German Empire by ... ten or twenty machines within the course of a few weeks'.

Here was characteristic Wallis fervour: he deserved full credit for conducting unfunded and unsupported research on the science of destroying large structures from the air, at a time when the RAF was institutionally indifferent to this vital issue. However, Wallis was as wrong as the 'bomber barons', and remained so throughout the war, in cherishing exaggerated expectations about what air power might achieve. He was as mistaken as Sir Arthur Harris, though from a different perspective, in believing that the RAF, or indeed the USAAF, could alone defeat Nazism, or even wreck the German economy. This objective was unattainable, regardless of which targeting policy

the air forces espoused, or what bombs he might devise for
Allied aircraft to carry. Yet Wallis cherished one remarkable
idea, that would secure his place in history.

2 GESTATION

Barnes Wallis knew nothing about the Air Staff's exploration
of targeting dams when, early in the war, he himself began
studying the vulnerabilities of German power supplies, and
explicitly of hydro-electric plants, during spare hours snatched
from his 'proper' work on a projected high-altitude Wellington,
and later the Windsor. He spent months considering the possibility of breaching dams with ten-ton bombs dropped by his
own proposed 'Victory' aircraft from an altitude of forty thousand feet – three times the operating height of contemporary
RAF 'heavies'. An early enthusiast for his ideas was Gp. Capt.
Fred Winterbotham, head of air intelligence at MI6, and a
pre-war pioneer of the exploitation of high-altitude aerial
photography. He was introduced to Wallis by a mutual friend,
City banker Leo D'Erlanger, who had endeared himself to the
engineer's children by once presenting them with a pink gram-
ophone. In February 1940 D'Erlanger brought the air intelligence officer to lunch at Effingham, thinking that Wallis and
Winterbotham had common interests. Winterbotham was
much taken with the cheerfully bustling Wallis household and
its noisy children, the exuberant piano-playing, the obviously
blissful partnership of his host and wife Molly.

Winterbotham was something of a charlatan, who played a
less important role both in the Second World War and the
evolution of Wallis's schemes than he himself later professed.
He was no fool, however, and like many intelligence officers
was a keen networker and intriguer. He invited Wallis to lunch

at the RAF Club in Piccadilly, and was persuaded by him to lobby the good and great about the Victory bomber, with a wingspan of 160 feet (against the Avro Lancaster's eventual 102 feet) and its accompanying 'earthquake' bomb. Desmond Morton, Winterbotham's old intelligence colleague, responded to this proposal from his new office in 10 Downing Street on 5 July 1940: 'My dear Fred ... The view held [here] is that such a project as you describe could not come to fruition until 1942, even if then.' This period was, of course, Britain's darkest of the Second World War; only by straining every sinew could the Ministry of Aircraft Production create a bare sufficiency of fighters, never mind a speculative giant bomber.

Nonetheless, through Winterbotham again, Wallis secured an audience with Lord Beaverbrook, Minister of Aircraft Production, at which he pressed his Victory project. The gnome-like tycoon seemed more interested in persuading his visitor to travel to America to explore pressurised aircraft cabins, but their meeting yielded one positive result: it enabled Wallis to secure access to the government research facility at the former Road Research Laboratory at Harmondsworth, just west of London, together with the Building Research Station near Watford in Hertfordshire.

In August 1940 Wallis began tests related to the projected deep-penetration bomb, for which he was also admitted to the wind tunnel at Teddington's National Physical Laboratory. In retrospect it seems astonishing, and yet also a triumph of official imagination, that even while Britain faced its darkest days, and Fighter Command was challenging the Luftwaffe against odds, a 'boffin' was able to undertake such futuristic research almost literally on the ground beneath which the Battle of Britain was being fought. From October onwards Wallis attended a series of meetings with the Ministry of Aircraft

Production's Air Vice-Marshal Francis Linnell, Controller of Research and Development, and Dr David Pye, the MAP's director of scientific research, together with his deputy Ben Lockspeiser. The last, especially, would play a role in the Chastise saga which continued until the day the operation was launched.

In November the RRL's Dr Norman Davey began construction of a 1:50 scale model of the Möhne, across a small stream in secluded woodland at the BRS in Hertfordshire. This project reflected the interest not merely of Wallis, but of the RAF's most senior officers, who had identified the dam as a target. At the same period Wallis was granted access to the Air Ministry's 1939 research on the Möhne, emphasising the fact that he and the uniformed planners had been thinking along parallel lines. This made Wallis all the more irritated that so many bureaucratic obstacles were placed in the way of what seemed to him an obvious war-winner. In November 1940 also, he wrote a testy note to AVM Arthur Tedder, then serving at the MAP: 'As a result of the continuing opposition that we have met, it has been necessary to resort to these laborious and long-winded experiments, in order to prove that what I suggested last July [destroying targets with deep-penetration bombs] can in reality be done.'

Norman Davey's team employed technical data on the Möhne dam's construction published at the time of its opening in 1913, though the Watford modellers somewhat distorted their own outcome by treating metres as yards in the scaling exercise. Hundreds of thousands of hand-cast mortar blocks were made and laid in the freezing conditions that prevailed through that winter. The model was completed on 22 January 1941, and explosive tests began a few days later. The first results of these were felt by nearby 'Dig for Victory' vegetable allotment-holders, who found their plots at Garston suddenly flooded by an inexplicable onrush of water. This also bewildered the BRS testers, because while their dam was damaged by successive explosions, it was not completely breached.

In March 1941 Wallis circulated a long paper entitled 'A Note on Methods of Attacking the Axis Powers', in which he wrote about water and coal seams as targets. Such natural resources, he observed, had the great merit that they could not be moved or dispersed: 'If their destruction or paralysis can be accomplished, THEY OFFER A MEANS OF RENDERING THE ENEMY UTTERLY INCAPABLE OF CONTINUING TO PROSECUTE THE WAR.' He distributed a hundred copies of this paper, with its extravagant predictions, to his aviation contacts – several journalists received it, together with four Americans and Frederick Lindemann, soon to become Lord Cherwell. Wallis's daughter later remarked on her father's carelessness about security: 'I can hear him now, describing to a friend some interesting feature of his work, laughing, "Frightfully secret, my dear fellow."'

W/Cdr. Sydney Bufton, an officer with operational experience over Germany who had recently become deputy director of Bomber Operations at the Air Ministry, was sufficiently interested to visit Wallis in his office at Burhill Golf Club, near Weybridge, where the design team found a wartime home after the Vickers plant was bombed. A dams sub-committee was formed at the Ministry of Aircraft Production, which in the following month discussed the Möhne as an important target. Initial calculations suggested that a bomb weighing twelve tons would be required to destroy it.

On 11 April 1941, David Pye of the Road Research Laboratory convened a meeting about Wallis's various advanced weapons concepts with the AAD – Aerial Attack on

Dams – Advisory Committee, which was also attended by the great scientific civil servant Sir Henry Tizard. At this it was concluded that the science of Wallis's ideas about destroying dams seemed sound: the intractable problem persisted, however, of devising a means of delivering to Germany a weapon such as might create the impact that he sought. This was no mere detail, but the core of the issue with which the Vickers engineer and the many technicians associated with his project would wrestle for the next two years.

Their progress was impeded, not by a mindless bureaucracy, but instead by practical difficulties which had to be addressed with severely constrained resources. Wallis scarcely helped his own case by arguing as if he, and he alone, held the key to winning the war. This was a vice to which bigger men were also prone. In September 1941 Churchill rebuked Portal, the chief of air staff, for submitting to him a paper which promised that if Britain built four thousand heavy bombers, the RAF could crush the Nazis within six months, without need for assistance from the other two services.

The prime minister responded in one of his most brilliant memoranda: 'Everything is being done to create the bombing force on the largest possible scale ... I deprecate, however, placing unbounded confidence in the means of attack, and still more expressing that confidence in terms of arithmetic ... Even if all the towns of Germany were rendered largely uninhabitable, it does not follow that the military control would be weakened or even that war industry could not be carried on ... The Air Staff would make a mistake to put their claim too high ... It may well be that German morale will crack, and that our bombing will play a very important part in bringing the result about. But all things are always on the move simultaneously ... One has to do the best one can, but he is an unwise man who thinks there is any certain method of winning this war, or indeed any other war between equals in strength. The only plan is to persevere.'

The prime minister would assuredly have said the same wise things to Barnes Wallis, had he been party to the correspondence about his putative wonder-weapons. On 21 May 1941 the engineer received a letter from Sir Henry Tizard, telling him that his ideas for both the Victory bomber and the deep-penetration bomb had been rejected by the Air Staff. Wallis was distraught. His fortunes had reached their lowest wartime ebb.

What followed, albeit painfully slowly in Wallis's eyes, reflected an important contradiction about the conduct of the Second World War. As a fighting force, man for man, from beginning to end the Wehrmacht showed itself more profes-sionally skilful than either the British or American armies. Yet the Western Allies nonetheless contrived to make better war than did the Axis powers. An important part of the reason for this was that they empowered many of the brightest people in their societies to deploy their talents, with an imagination which the dictatorships never matched. The codebreakers of the US Navy's Op20G and the US Army's Arlington Hall, together with Britain's Bletchley Park, provided conspicuous examples of this phenomenon. So, too, did a host of projects commissioned and undertaken by scientists and engineers on both sides of the Atlantic.

Although Barnes Wallis's Big Plane, Big Bomb proposals had been formally rejected in May 1941, he nonetheless persuaded the MAP's David Pye that he should retain access to government facilities, to continue his experiments on the ballistics of dam-breaking. Through that autumn tests contin-ued, to determine the necessary weight of explosives, and the conditions in which they must be detonated, to contrive breaches in huge structures.

It was an elaborately formal age. Many of the papers in what became a mountainous correspondence between Whitehall's civilian and service departments about the engineer's infernal machines began as did this one to an under-secretary of state: 'Sir, I have the honour to state that consideration has again been given recently to the possibility of breaching one or more of the important canals in North West Germany.' The engineer concerned was referred to 'as Mr B.N. Wallis of Vickers'. The writer signed himself 'your obedient servant'.

From the £2,000 budget then allocated to Wallis's activities by the MAP, money was found to buy from Birmingham City Council a small dam at Nant-y-Gro in Powys, North Wales, rendered redundant by the construction of a larger replacement. A key figure in the experiments that followed was Arthur Collins, a scientific officer in Harmondsworth's 'Concrete Section', who made a breakthrough. For years it had been assumed, not least by Barnes Wallis, that an enormous explosive charge would be necessary to destroy a dam such as the Möhne. Yet experiments convinced Collins, who in turn persuaded Wallis, that a relatively small charge might achieve a wholly disproportionate result if it was detonated sub-aqueously and close to the target, using a timer or a hydrostatic pistol: it could thus harness the power of the water mass to channel the force of the blast. Here was the phenomenon identified as a threat back in 1939 by the German official responsible for his country's north-western dams. Both Collins and Wallis became increasingly fascinated by the physics of explosions, and especially by the scope for harnessing the power of water, and indeed of earth, dramatically to increase the impact of underwater or underground explosions

– the 'conservation of suspended energy' that would eventually make possible Operation Chastise.

In the course of 1941 and 1942, Wallis pursued enquiries about Germany's dams through patent agents in Chancery Lane, and about hydro-electric control mechanisms via an engineering firm in Kilmarnock. In April 1942 – Holy Week, as it happened – experiments assisted by his children, using marbles projected into an old galvanised washtub on the terrace outside his home at Effingham, shifted his attention from deep-penetration 'earthquake' charges towards the notion of much smaller spherical bombs, bowled – in cricketing parlance – or ricocheted – to use Wallis's original choice of word – towards German dam walls. Here, he was thinking in a fashion not dissimilar from Finch-Noyes and Pemberton-Billing. He envisaged two related, but different weapons: a larger model for attacking dams, later codenamed 'Upkeep', as it will hereafter for convenience be called; and a smaller version, to be codenamed 'Highball', for use against shipping.

Sir Charles Craven, a former Royal Navy submarine officer who was now chairman of Vickers, did not explicitly bar Wallis's spare-time work on futuristic weapons. He emphasised, however, that it must not interfere with the engineer's day job, developing the Windsor bomber. In post-war evidence to the Royal Commission on Awards to Inventors, Wallis stated that 'the inception of the [bouncing bomb] was the result of private experiment and work outside the scope of his normal employment and that this work was carried out against the wishes of his employers'. He subsequently expanded on this theme, saying that 'had he not persisted in his efforts to interest the authorities in the face of continued discouragement and even contrary to the wishes of his own Directors, the attack on the dams would never have been made'. In the narrative that follows, it should not be forgotten that, until the last stage of the development of Wallis's revolutionary weapons, his work on them represented, in the stern view of his employers, a spare-time indulgence.

3 FIRST BOUNCES

In the late spring of 1942, Barnes Wallis reported to the MAP and the Air Ministry that he believed he could overcome a critical problem – accurately to deliver a charge from a fast-moving bomber against a target protected with anti-torpedo nets – by bouncing a bomb across the water in the fashion he had explored with marbles on his terrace at Effingham. Moreover, a century and a half earlier Vice-Admiral Horatio Nelson and his fellow Royal Navy commanders had shown the way, exploiting the technique of bouncing cannonballs across the sea to pummel French warships. At the end of May, Wallis set off with his secretary, former British ladies' rowing champion Amy Gentry, for Silvermere Lake near Cobham to test the potential of using a catapult, much more sophisticated than a child's toy, to bounce small projectiles down a test tank. In the course of these experiments they found that, if a golf-ball-sized object was backspun on release, it would 'ricochet' far more vigorously. Vickers' experimental manager George Edwards, a keen cricketer, later claimed credit for this idea, but the evidence suggests that Wallis developed it himself, and merely had later conversations about it with Edwards.

The eventual form of Upkeep was that of a large, cylindrical naval depth-charge. Until late April 1943, however, Wallis envisaged its shape as almost or absolutely spherical, the huge canister containing the charge being encased in an outer shell of wood. It was also at times described as a mine, which became part of its cover story in official correspondence and later news coverage. Since legend, however, knows the dam-busting weapon as a bomb, that is how it will continue to be described in this narrative.

Wallis told Fred Winterbotham that he saw every reason to believe that the new weapon's destructive principles would prove as applicable to enemy shipping as to dams, locks and suchlike. Thus, on 22 April 1942 Winterbotham accompanied the engineer to discuss the project with Professor Pat Blackett, the exceptionally enlightened physicist who was scientific adviser to the Admiralty. Blackett, in turn, lobbied Tizard, who despite his opposition to Wallis's big-bomb project a year earlier was now sufficiently excited to visit him at Burhill on the 23rd. Tizard thereafter supported Wallis's request for access to two experimental ship tanks at the National Physical Laboratory at Teddington, where he began tests in June which continued over twenty-two days, at intervals until September. If the pace of progress appears slow, it must be remembered that Britain was still conducting its war effort on desperately short commons, while Wallis was earning his bread working on the Windsor bomber.

Although the Royal Navy was perhaps Britain's most successful armed service of the war, the Fleet Air Arm was its least impressive branch. Despite the much-trumpeted success of a November 1940 torpedo attack on Italian capital ships in their anchorage at Taranto, carried out by antiquated Swordfish biplanes, thereafter British naval aircraft enjoyed few successes. Churchill more than once acidly enquired why the Japanese seemed much better at torpedo-bombing than was Britain's senior service. Admirals were thus immediately attracted to a new technology which might make the Fleet Air Arm less ineffectual. For months after Wallis's 'bouncing bomb' was first mooted, the RAF sustained institutional scepticism; sailors did more than airmen to keep the concept alive.

Tizard himself attended some tests at Teddington, as did Rear-Admiral Edward de Faye Renouf, a former torpedo specialist who was now the Admiralty's director of special weapons. Renouf and several of his staff watched a demonstration in which a two-inch sphere was catapulted down a tank, bouncing along the water until it struck the side of a wax model battleship and rolled down beneath its hull. The admiral, a gifted officer recently recovered from a nervous breakdown after a succession of terrifying experiences while commanding a cruiser squadron in the Mediterranean, urged Sir Charles Craven of Vickers to give priority to Wallis's weapons research. Renouf envisaged a projectile that might be released from the new twin-engined Mosquito light bomber.

That month, May 1942, Wallis produced a new paper incorporating all this research, entitled 'Spherical Bomb, Surface Torpedo'. His thinking still focused entirely on round weapons, described in a note from Winterbotham to the Ministry of Production as 'rota-mines'. Wallis's paper cited earlier work by a German scientist, and also showed that for a bomb to get close enough to a dam to enable the principle of 'Conservation of Suspended Energy' to work, it needed to impact upon the water almost horizontally, at an angle of incidence of less than seven degrees, which meant that it must be dropped from an aircraft flying very low indeed: at that time, 150–250 feet seemed appropriate. Wallis envisaged its release from a range of around twelve hundred yards, to allow time for the attacking pilot to turn away and escape before flying headlong over the target and its defences. Not until months later was a requirement accepted for the aircraft to carry its bomb much closer, and thereafter to overfly the objective.

In a further demonstration of the validity of Churchill's observation that 'All things are always on the move simultaneously,' at the Road Research Laboratory Arthur Collins had meanwhile been conducting a succession of tests on two 1:10 scale models of the Nant-y-Gro dam. On 10 May 1942 Wallis and his wife Molly travelled to Wales with Collins's team to witness experiments on the full-sized dam. These established that if an explosion took place at any significant distance from its wall, the blast was too weak to precipitate a fracture. Collins wrote: 'A solution to the problem was, however, found almost by chance shortly afterwards.' His team needed to remove one of the damaged scale models at Harmondsworth, and used a contact charge to shift the concrete. The result was devastation, on a scale unmatched by any 'near-miss'.

Further tests confirmed the result, and on 16 July Wallis received an invitation to attend a full-scale demonstration a week later. He was nettled by the short notice, and warned a little pompously that he was working under such pressure – presumably on the Windsor bomber – that he would probably be unable to get away. Nonetheless, he was present at Nant-y-Gro when, on the 24th, army engineers blew a 279-lb charge of which the effects were filmed with high-speed cameras brought to North Wales from the Royal Aircraft Establishment at Farnborough. The test explosion proved a triumph, blasting a breach in a masonry construct that was, for practical purposes, a small-scale version of a German dam.

In the following month, Collins submitted a report which concluded that if a charge weighing around 7,500 lb was exploded at a depth of thirty feet against the wall of a dam such as the Möhne, it should be capable of achieving a breach. Such a weapon would not require the creation of a new bomber to carry it, but was within the lifting capabilities of the new Avro

Lancaster, subject to appropriate modifications. Thus, suddenly, the most intractable obstacle to an attack on Germany's masonry dams was removed: it seemed feasible – in theory at least – to convey to the target sufficient explosive to destroy it. Credit for the principal scientific achievements that made possible Operation Chastise should rightfully be shared between Collins, who resolved the challenge posed by the physics of destroying a vast man-made structure, and Wallis, who conceived a technique whereby the necessary charge might be laid from the air with the exactitude indispensable to success.

In the late summer of 1942, a situation obtained wherein Barnes Wallis had devised a revolutionary weapon, of which the scientific principles were agreed by most of the experts who studied them to be sound. The Royal Navy was excited about its possibilities for use by the Fleet Air Arm. Widespread scepticism nonetheless persisted, shared by MAP's David Pye and his deputy, Ben Lockspeiser, about whether the resources could be justified to pursue a speculative technology that could only be used over water, and which demanded superhuman courage and skill from aircrew who would have to launch it against an enemy. Moreover, every aircraft which carried such a bomb would require expensive modification.

Such reservations were fully justified. Lockspeiser wrote to Tizard on 16 June: 'It is quite impractical and uneconomic to modify our bombers in large numbers for the special purpose of carrying any particular bomb.' Nonetheless the Admiralty's enthusiasm, and the uneasy acquiescence of MAP's AVM Linnell, sufficed to secure a request for Vickers to fit a Wellington twin-engined bomber to carry a prototype Wallis bomb, of which on 22 July an order for twelve examples was placed with the Oxley Engineering Company. On 25 August Wallis attended a meeting at MAP at which arrangements were agreed for a series of trials to be conducted a month later, at Chesil Beach in Dorset.

It is striking to notice, at this stage, two camps in the service ministries and the defence scientific community about the whole project. One faction believed that Wallis's weapons were fanciful; would never work. The other cherished wildly over-optimistic fantasies concerning their war-winning potential. Fred Winterbotham wrote to the parliamentary secretary at the Ministry of Production on 14 September 1942, speculating about what Wallis's bombs might achieve: 'If this new weapon is intelligently used, e.g. for simultaneous attacks on all German capital ships and main hydro-electric power dams, there is little doubt but that Italy could be brought to a complete standstill and that industry in Germany would be so crippled as to have a decisive effect on the duration of the war ... To attain this result much preparation and careful planning are clearly required and meanwhile I repeat *nothing is being done*.' Here was a manifestation of a British yearning, characteristic of its time and place, for some dramatic stroke that might side-step battlefield slaughter and bring the war to an early closure. Winterbotham's note took no heed of the Alpine difficulties in the way of his fantasy, prominent among them that its fulfil-ment would require hundreds of bombers to be modified to carry Wallis's weapons. Meanwhile its expectations about what these, or for that matter any, bombs might do to the Axis drifted into fairyland.

That autumn of 1942, the bomb project languished. Oxley Engineering experienced difficulties in constructing the test weapons, and in October Wallis was kept at home for several days by illness. Only on 2 December did he at last board a modified Wellington, piloted by veteran test pilot Mutt Summers – the very same who had parachuted more than a decade earlier from one of the engineer's less successful proto-types – for a trial of the backspin technology. It worked, though no test bombs were dropped.

Two days later, on the afternoon of the 4th, the Wellington took off for Dorset, where on Chesil Beach a camera crew waited to record the trial bomb-dropping. The first two tests, with non-explosive fillings, resulted in the spheres bursting on impact. When a subsequent succession of droppings took place, Wallis watched from the shore. Outcomes suggested that the technologies for releasing the bomb from an aircraft, and for its subsequent bouncing progress, were viable. Yet repeated collisions with the sea at speeds of over 200 mph imposed enormous stresses on the projectile during its bounc-ing progress. Half-sized prototypes disintegrated. Their beget-ter undertook modifications and adaptations, still convinced that the principles of his creation – or rather, his instrument of destruction – were sound.

Following these further developments, Wednesday, 20 January 1943 found Wallis ensconced in Weymouth's Gloucester Hotel, from which he wrote to Molly: 'I do wish you could come & share this lovely room. It would be just perfect with you here. If you could come tomorrow by the mid-day train, do … Now we are scheduled to start [tests] at 10 a.m. … so I must go to bed. All my love little sweetheart, and come if you can …'

She came. Almost two decades of marriage had done noth-ing to cool the couple's passionate romance. Molly, still only thirty-seven, was now responsible for six children, including a niece and nephew whose parents had been killed in the blitz. This brood was surrendered to their nanny while she set forth for Dorset. On 24 January she wrote to the children from Weymouth about her visit to their father:

> … A lovely time it's been. This morning I drove out with him and the others to the nearest point I was allowed [to the bomb tests] & then I got out and walked back. 8 miles of lovely road, high up with the Downs one side & the sea to other. It was sunny & clear. I did enjoy it. Of course it's quite mad that they shouldn't let me watch the proceedings seeing as I've lived with it since last February & probably know more about it than anyone save Barnes. But it's quite true – policemen do bob up and turn you away. I suppose the others would say 'If that wife why not my wife' little knowing what a very special wife this one is.
>
> Darling Barnes. You should see him among these admirals and air vice marshals patiently explaining and describing to them & they drinking it all in – or trying to. And he's so quiet and un-assuming none of them could imagine what pain & labour it's been. How he's got up in the middle of the night to go up to the study & work summat out. No wonder he looks drawn and tired. I suppose if he were a self-advertiser he'd have been Sir Barnes in the New Year Honours. Oh well, it'd have been a nuisance. But it's an exciting life & no mistake.

The contrast seems extraordinarily moving between the private domesticity and indeed passion within the Wallis family, and the devastating public purposes which its principal was pursuing. Some might find it repugnant, in the light of what later befell Upkeep's victims, but it must never be forgotten that Barnes Wallis's country was engaged in a war of national survival against one of history's most evil forces. The engineer was straining every sinew, and all his own astonishing gifts, to assist the Allied cause. During that 24 January trial in Dorset which Molly Barnes was not permitted to witness, one dummy bomb achieved thirteen bounces. Next day, another managed twenty. Wooden test spheres, dropped from between eighty and 145 feet, travelled 1,315 yards across the water. Barnes once wrote down for Molly the closing lines of Tennyson's 'Ulysses', among his favourite poems: 'One equal temper of heroic hearts,/Made weak by time and fate, but strong in will/ To strive, to seek, to find, and not to yield.'

Returning to London, the white-haired evangelist now strode through the corridors of ministries proudly clutching a can of 35mm film, showing his weapon skimming the sea. Eagerly, he awaited authorisation to continue with development of both Upkeep – the dam-bursting version – and Highball, the smaller naval bomb. He chafed for a swift commitment, because for optimum effect an attack on the German dams needed to take place in May, when water levels in the reservoirs were at their maximum height after winter rains and snows.

Instead, however, on 12 February 1943 a blow descended. Ben Lockspeiser told Wallis that AVM Linnell had become concerned that the engineer's labours on his bombs – nobody used the word obsession, but this was obviously in many service minds – was impeding his 'proper' work, development of the Windsor bomber. Linnell did not explicitly oppose the bomb scheme – he was too canny a service politician for that. He merely reported on its speculative character to AVM Ralph Sorley, assistant chief of air staff for Technical Requirements, further emphasising the drain that the project was imposing on resources. Once again, it seems worthy of emphasis that Linnell did not thus play the part of a myopic senior officer flying a desk, but was instead assessing Wallis's project from the viewpoint of a department besieged by competing demands for facilities to develop new aircraft and weapons systems. Meanwhile Syd Bufton, deputy director of bomber operations, told a 13 February meeting at the Air Ministry that he, as an experienced operational pilot, considered it impracticable to drop Wallis's bombs in darkness, at low level over enemy territory.

Gp. Capt. Sam Elworthy, a Bomber Command staff officer who attended the same meeting, was charged with reporting on its findings to Harris's headquarters at High Wycombe, which he did on the following day. The consequence was a note on the 'bouncing bomb' drafted by AVM Robert 'Sandy' Saundby, senior air staff officer to his chieftain. This, in turn, prompted the C-in-C to scribble one of his most famous, or notorious, judgements of the war: 'This is tripe of the wildest description … There is not the smallest chance of it working.' And much more of the same.

Wallis now wrote an anguished personal note to Fred Winterbotham, expressing his frustration: 'We have just worked out some of our results from the last experiment at Chesil Beach, and are getting ranges nearly twice those which would be forecast from the water tank, that is, with a Wellington flying at about 300 miles an hour and dropping from an altitude of 200 feet, we have registered a range of *exactly three-quarters of a mile!!*' He said that the problems of constructing prototypes to be carried by a heavy bomber would be easily solved. 'It follows that sufficient bombs for the Lancaster experiment (if, say, thirty machines were to be used, to destroy simultaneously five dams, that is, six machines per dam to make certain of doing it) can be completed within two or three weeks.' He added that modifying the Lancasters would be a far more time-consuming process than manufacturing the weapons, and concluded 'Yours in great haste,' adding a hand-written scrawl: 'Help, oh help.' It was characteristic of the strand of naïveté in Wallis that in calculating the number of aircraft needed to destroy five dams in enemy territory he was heedless of the possibility that the German defences might remove from the reckoning some, if not all, of the attackers.

Winterbotham responded by writing on 16 February to AVM Frank Inglis, assistant chief of air staff for Intelligence. He extravagantly described the bouncing bomb as an invention 'for which I was partly responsible'. He asserted that the chief of combined operations and the prime minister were enthusiastic, though there is no shred of documentary evidence of Churchill's involvement at any stage. He then employed an argument often advanced by estate agents: if the Royal Air Force did not snap up this opportunity, the Royal Navy was eager to do so: 'My fear is that a new and formidable strategic weapon will be spoiled by premature use against a few ships, instead of being developed and used in a properly coordinated plan.' He urged ensuring that the chief of air staff was briefed, before it was too late.

Despite Harris's attitude, and airmen's continuing doubts about the tactical feasibility, on Monday, 15 February, Gp. Capt. Syd Bufton chaired a further meeting at the Air Ministry, attended by Elworthy, Wallis, Mutt Summers and others, in fulfilment of an Air Staff instruction 'to investigate the whole [dams] operational project'. Wallis delivered a superbly eloquent sales patter. Upkeep, he said, could be released from a height of 250 feet, at a speed of around 250 mph, at a distance from the target of between three-eighths and three-quarters of a mile. Responding to Elworthy's concern, expressed on behalf of Bomber Command, about a diversion of precious Lancasters for modification, he said that only one aircraft would be needed for full-scale trials, while those used for an Upkeep attack could be restored to normal operational mode within twenty-four hours. He suggested that while the Möhne dam was the most prominent target suited to Upkeep, the Eder – forty-five miles east-south-eastwards – was also vulnerable.

Most of this was debatable, and some of it flatly wrong. Nobody at the meeting pointed out that even if the Eder represented a suitable target for bouncing bombs, it was un-related to the Ruhr water system, which was supposedly the strategic objective. The aircraft to carry Wallis's weapons did not require mere modification, but would instead need fuse-lages purpose-built by Avro, and could not thereafter be read-ily returned to Main Force duty. Wallis's persistence emphasised his gifts as a street-fighter. Where his professional passions were engaged, he was a much less gentle, more ruth-less man than was sometimes supposed by those who met him casually. On this occasion, his reputation and conviction carried the day. Bufton changed his mind, renouncing the disbelief he had expressed on 13 February to report in the name of the committee: 'It was agreed that the operation offered a very good chance of success, and that the weapons and necessary parts for modification should be prepared for thirty aircraft.' It was thought that as long as the attack took place before the end of June, reservoir levels should be high enough to create massive flooding.

Bufton told AVM Norman Bottomley, assistant chief of air staff for Operations, 'the prospects offered by this new weapon fully justify our pressing on with development as quickly as possible'. Bottomley, who would play an important role in securing the final commitment to the dams raid, was a veteran networker within the corridors of power. Syd Bufton said of him with wry respect: 'Nobody could play the Air Ministry

organ as skilfully as Norman.' It was Wallis's additional good
fortune that Bufton and Elworthy – a thirty-one-year-old New
Zealander of outstanding abilities who eventually became
head of the RAF – were original thinkers, open to new ideas in
a fashion that Harris was not. They grasped the terrific theatrical impact that the dams' destruction would make, surely
greater than that of yet another assault on German cities.
Churchill once said grumpily, 'I'm sick of these raids on
Cologne,' to which Sir Arthur Harris's riposte – 'So are the
people of Cologne!' – was not wholly convincing.

A weakness of the debate about Upkeep, however, was that
it focused overwhelmingly on the feasibility of constructing
and dropping the bombs; much less on the vulnerabilities of
the water systems of western Germany, the Ruhr in particular.
Throughout the Second World War, intelligence about the
German economy and industries remained a weakness in
Western Allied warmaking, and explicitly in the conduct of the
bomber offensive.

Just three days after the Air Ministry meeting, on 18
February, following a telephone conversation with Linnell of
MAP, who remained a sceptic, Harris wrote a testy note to
Portal, his chief, head of the Royal Air Force. Linnell had told
him, he said, 'that all sorts of enthusiasts and panacea-
merchants are now coming round MAP suggesting the taking
of about thirty Lancasters off the line to rig them up for this
weapon, when the weapon itself exists so far only within the
imagination of those who conceived it. I cannot too strongly
deprecate any diversion of Lancasters at this critical moment
in our affairs.' Wallis's bomb, in Harris's view, 'is just about the
maddest proposition … that we have yet come across … The
job of rotating some 1,200 pounds [sic] of material at 500 rpm
on an aircraft is in itself fraught with difficulty.'

But Wallis had acquired supporters more powerful even than Harris. After a screening of a new batch of films of his tests before audiences that included Portal, First Sea Lord Admiral Sir Dudley Pound and Vickers chief Sir Charles Craven, Pound threw his weight behind the naval version of Wallis's mine: 'The potential value of Highball is so great,' he minuted on 27 February, '… that not only should the trials be given the highest priority, but their complete success should be *assumed* now.'

It is hard to overstate the stress under which the bomb's begetter existed in those days. He was still spending many hours on the design of the Windsor, Air Ministry specification B.3/42. In the mind of Sir Charles Craven, this was much more important than Upkeep and Highball: contracts for a big new bomber promised immense rewards for Vickers, contrasted with those to be gained from building a few bombs. Moreover, confusingly for Wallis and for the entire Whitehall hierarchy, Craven was intermittently seconded to assist and work in the Ministry, so that it was sometimes unclear to all concerned whether he spoke as the engineer's employer, or as the voice of officialdom. On 18 February, Wallis worked until 7.45 p.m. at the National Physical Laboratory. Next morning, he met the Admiralty's director of weapons development, then at 2.30 p.m. saw MAP officials to discuss unspecified aircraft de-icing problems. At 4 p.m. he was back at Vickers, where at 5.30 p.m. there was another screening of his bomb-test films, following which he drove to Dorking with Admiral Renouf. On the next morning, a Saturday, he worked in his office at Burhill, then attended more meetings in the afternoon. On Sunday he was confined to his home at Effingham with a migraine, such as he often and unsurprisingly succumbed to.

Next day, Monday the 22nd, he drove with Mutt Summers to High Wycombe for a personal audience with Sir Arthur Harris. Sam Elworthy claimed credit for persuading the C-in-C to meet the engineer. He wrote to Harris after the war, saying emolliently that 'your scepticism of what seemed just another crazy idea was certainly shared by your staff'. But the clever group-captain had been impressed by what he heard about Upkeep – and he also saw which way the wind was blowing at the Air Ministry.

Harris was no fool. For all his bombast, he grudgingly acknowledged that he had masters who must sometimes be appeased. He knew that Portal had authorised the modification of three Lancasters to carry Upkeep. While the words 'whether the Commander-in-Chief of Bomber Command likes it or not' were never articulated, they were understood. The Chief of the Air Staff wrote to Harris on 19 February: 'As you know, I have the greatest respect for your opinion on all technical and operational matters, and I agree with you that it is quite possible that the Highball and Upkeep projects may come to nothing. Nevertheless, I do not feel inclined to refuse Air Staff interest in these weapons.'

That morning of the 22nd at High Wycombe, Wallis was subjected to a predictable barrage of invective: 'What is it you want? My boys' lives are too precious to be wasted on your crazy notions.' Yet it is unlikely that Harris would have received Wallis at all had he not already recognised that he would have to give way, and provide resources for an operational trial of Upkeep. Having viewed the films, he professed grudging interest.

Wallis succumbed to a brief surge of optimism. This was shattered, however, on his return to Weybridge. He received an order to present himself immediately at the London office of

Vickers, his employers, for an audience with the company's chairman. Craven, without inviting his visitor to sit down, declared brusquely that MAP's Linnell had complained that Wallis had become a nuisance; that his bouncing bombs had become a serious impediment to the vastly more important Windsor project. The air marshal had explicitly demanded that Craven call a halt to Wallis's 'dams nonsense'.

This was what the Vickers chairman now did. A shouting match followed, in which the designer offered his resignation, and Craven shouted 'Mutiny!' They parted on terms of mutual acrimony. Moreover, while Wallis was not often a grudge-bearer, he never forgave AVM Linnell for the part he played in attempting to kill off Upkeep. He went home despondent to Effingham, sincerely determined upon resignation, as was scarcely surprising after the humiliation he had suffered. Craven, whose responsibility was to Vickers, can scarcely be blamed for his behaviour, after being told by the Ministry of Aircraft Production – upon whose goodwill his company depended for orders – that its chiefs were tired of his nagging, insistent assistant chief designer (structures). Why should such people as Linnell, Craven and indeed Harris have accepted at face value the workability of a new weapon which represented a marriage of technologies of extreme sophistica-tion with others of almost childlike simplicity, which when fitted to a Lancaster caused it to resemble a clumsy transport aircraft with an underslung load?

Yet Wallis knew that, whatever Craven said about the MAP's view of Upkeep, the Admiralty remained enthusiastic about Highball. On 26 February, by previous arrangement he drove to London to attend a meeting that was to be chaired by the now-detested Linnell, to discuss measures to improve the aerodynamics of what some described as 'the golf mine' –

because of its resemblance to the shape of a golf ball. After Wallis was told that Roy Chadwick of Avro, designer of the Lancaster, would also be attending, he understood that Craven had got things all wrong the previous day: the RAF had not abandoned Upkeep.

When the delayed meeting finally convened at 3 p.m. that Friday, in Linnell's office at MAP on London's Millbank, it was to receive tablets from on high. Sir Charles Portal was not only chief of air staff and a former C-in-C of Bomber Command; he had also been among the first enthusiasts for attacking Germany's dams. He was troubled by doubts about Sir Arthur Harris's obsession with destroying cities. His reservations were founded not upon moral scruples – no senior wartime airman admitted to those – but instead on uncertainty about its war-winning potential. Portal never summoned the courage to sack Harris, even in the winter of 1944–45, when his subordinate directly defied targeting orders. But the CAS was always at heart a proponent of attacking precision objectives, if means existed to make such a policy work. Now, Barnes Wallis promised to provide these; to make possible fulfilment of the RAF's 1937–38 dream, of an assault upon Germany's dams.

The array of brass assembled at the MAP on the afternoon of 26 February 1943 was told that the chief of air staff had given his assent, or rather had issued an order, to proceed with immediate development of Upkeep. Portal wrote: 'I think this is a good gamble.' The reaction of Sir Charles Craven, who was also present, is unrecorded. He must have felt privately foolish, if not furious, following his ugly dressing-down of his designer four days earlier. What the CAS demanded, however, Vickers must seek to provide.

Linnell, too, can scarcely have enjoyed announcing – against his own strong personal conviction – the decision to prioritise

Upkeep over the Windsor bomber. 'The requirement for bombs,' this MAP potentate now said, 'has been stated as one hundred and fifty to cover trials and operations' – one hundred and twenty were eventually made. It was further stated that studies of the German dams showed that 26 May – just three months ahead – was the latest date in 1943 on which they could plausibly be attacked. It would thus be necessary to build thirty 'Provisioning' Lancasters, as they had been codenamed, and to produce a sufficiency of bombs by 1 May, to provide reasonable time for aircrew training for the operation. The MAP's budget for research on Upkeep, which in August 1942 had been raised from £2,000 to £10,000, was now further increased to £15,000, and later again on 1 April to a princely £20,000.

While Portal made a personal commitment that enabled Upkeep to be unleashed, it deserves emphasis that he placed a relatively modest bet. Only air-power fantasists could suppose that a single squadron of Lancasters, a maximum of twenty bouncing bombs, would cripple the entire water system of the Ruhr, an aspiration demanding a much larger force even if Wallis's weapons were half-successful – almost no new weapons system in history has performed better than that. Yet a squadron was all that the Air Ministry would authorise, in the face of Harris's virulent hostility, a limited supply of aircraft and uncertainty about the viability of Upkeep. British warmakers have for centuries displayed a weakness for 'gesture strategy' – deploying a disproportionately small force as a means of displaying interest in fulfilment of a disproportionately large objective. Mass matters, however, to the success of all military operations, and in this case it would be lacking. Though the RAF neither then nor later admitted this, its commitment to an assault on Germany's dams was marginal, a tiny fraction of the forces that set forth upon almost nightly attempts to burn cities. To borrow a modern phrase, this would be a niche operation.

Wallis's commitment to Upkeep, by contrast, was total, and now confronted him with a dramatic challenge. At breakneck speed he must convert his theoretical concept into a viable bomb, while work was simultaneously rushed forward on building the modified Lancasters. Avro, the plane's manufacturers, agreed to fit the necessary electrical release gear, along with strongpoints where the bomb doors must be removed, and hydraulic power for backspin pulleys. Vickers would meanwhile make protruding retainer arms for the Upkeeps, to be attached to the strongpoints; a rotational driving mechanism; and the bombs themselves. All the latter work would be carried out at Weybridge, with Avro dispatching a team to work on the Vickers site. Wallis promised to provide working drawings of the latest version of Upkeep within ten days. Craven expressed concern about whether a revolutionary weapon of such size could be machined within the necessary time-frame.

Before the final decision, some minor obstacles had to be swept away. By coincidence Combined Operations, then headed by the frisky Lord Louis Mountbatten, was proposing an attack on the Möhne, which Mountbatten described as 'one of the great strategic targets'. Special Operations Executive also suggested an assault by parachute saboteurs. Both bodies' proposals were now quashed, fortunately for those who might have been charged with implementing them, in favour of what was designated Wallis's 'rolling bomb'. It would be the RAF which destroyed the dams of north-west Germany. Or nobody.

3

Command and Controversy

1 TARGETS

At the end of February 1943, more than six years after the RAF first discussed the feasibility of attacking Germany's dams, and over three years since the war began, air chiefs and engineers embarked upon a ten-week dash to launch the operation that was shortly thereafter codenamed Chastise. The strategic planning – above all, about target priorities – took place within the Air Ministry. This was gall and wormwood to Sir Arthur Harris, who regarded his own headquarters staff as the only proper people to arbitrate on such matters.

Wallis, in a paper headed 'Air Attack on Dams' that had been circulated to a range of interested parties on the secret list back in January, identified six plausible targets – the Möhne, Eder, Sorpe, Lister, Ennepe and Henne – of which the reservoirs held a combined total of almost nine thousand million cubic feet of water, while seven other, smaller dams retained just 423 million. Breaching the Möhne alone, he promised, would cause 'a disaster of the first magnitude' extending to the lower reaches of the Ruhr. He appears to have intended this judgement to describe solely the industrial consequences, and to have given no consideration one way or another to the inevitable human cost among civilians caught in the path of the intended deluges, whom air-raid shelters would avail little. Thereafter, to a remarkable degree the Air Ministry made its Chastise targeting decisions on the basis of the assessment advanced by Wallis, a professional engineer but an amateur analyst of Ruhr industries.

From the Directorate of Bomber Operations, Gp. Capt. Syd Bufton* wrote to AVM Norman Bottomley, Portal's deputy, urging the primacy of the Möhne over the Eder – also mentioned by Wallis: 'The former is much more important, and tactically the more suitable. We should, therefore, ensure that an adequate effort is devoted to this objective before considering the possibility of an attack upon the Eder.' He added in a covering note: 'I feel a lot of time would be wasted if we go to the lengths of making a minute examination of all possible targets.'

Intelligence was a fundamental weakness of Britain's strategic air offensive. Bomber Command's decisions about targets were often made on the basis of their accessibility or vulnerability to attack, which had earlier in the war prompted an emphasis on relatively easily-located coastal objectives. Before an operation, immense effort was expended upon fixing routes, providing electronic aids and counter-measures, deciding bomb and fuel loads, and nominating diversionary landing fields. By contrast, examination of targets' economic significance was often as superficial as post-attack damage assessment. In planning for Chastise, the Möhne and the Eder were prioritised because both were masonry structures, and thus most likely to succumb to Wallis's bombs. Yet while the Möhne was a hub of the Ruhr industrial water system, the Eder was

---

* In March Bufton was promoted from deputy to director, or DBO.

unrelated to it. At an early stage of the war, the Ministry of Economic Warfare's experts had concluded that if the Möhne and the nearby Sorpe could both be breached, the effect on Ruhr water supplies might well be catastrophic. If only one was destroyed, however, stated the MEW, the Nazis could probably contain the threat to Ruhr industrial activity.

The Sorpe was an immensely thick earthen dam, which sloped steeply on the reservoir side in a fashion which ensured that a mine which was bounced up its lake towards the wall must roll away backwards on impact. On 18 March an Air Ministry ad hoc Chastise committee, chaired by Bottomley, agreed that the Sorpe's construction thus ruled it out as a target. It was also decided that a strike with Highballs, carried by Mosquitoes of the RAF's Coastal Command, should be attempted more or less simultaneously against *Tirpitz* or another German battleship in the Norwegian fjords. Bottomley reported to the chiefs of staff, outlining progress: 'the speed with which [Highball and Upkeep] have been developed has been so high and the time available to complete them before the required date is so short, that there is a considerable element of gamble'.

A fierce division of opinion about scheduling now emerged between the RAF and the Royal Navy. The airmen had become committed to attacking the dams before the end of May. The sailors, however, recognised that the first use of bouncing bombs, whether Upkeeps or Highballs, could also prove the last, because the Germans would adopt counter-measures to protect vulnerable watery targets. Thus, the navy wanted Chastise deferred until Coastal Command was ready to attack German warships with Highballs, an operation codenamed Servant. It was acknowledged that the tri-service chiefs of staff might have to be asked to arbitrate on this knotty issue.

The airmen meanwhile reached an irrational compromise about the Sorpe dam, such as was typical of the entire bomber offensive. While acknowledging that it was unsuitable as an objective for bouncing bombs, it was restored to the target list because of its agreed strategic importance. Wallis thought that four or five of his Upkeeps might breach it without being bounced. He cited the precedent of the 1864 natural collapse of the Dale Dyke dam near Sheffield, which destroyed six hundred homes and killed at least 240 people. An earth dam with a concrete core, he claimed, would become 'practically self-destroying if a substantial leak can be established within the water-tight core'.

This may have sounded persuasive, but it was in fact ill-founded: since the Dale Dyke dam had a clay rather than a cement core, it was not comparable with the Sorpe. The planners eventually decided that the Wallis bombs used against the Sorpe must be dropped following a lateral, overland approach, without backspin, to sink and explode on time fuses rather than by pressure upon a hydrostatic pistol. It seems remarkable that Wallis or anyone else supposed that Upkeeps would thus achieve their purpose-designed ballistic effect any more than might any other explosive charge of similar size – in other words, a conventional bomb. It is hard not to suspect that the engineer asserted the plausibility of destroying the Sorpe because he feared that if he did not do so, the entire commitment to Chastise would once more be thrust into doubt because of that dam's strategic centrality.

On 25 March another meeting of the Air Ministry committee, attended by an array of brass representing both the Royal Navy and the RAF, including Saundby from Bomber Command, was told that construction of the 'Type 464 Provisioning' Lancaster variants was proceeding on schedule.

Sixteen Mosquitoes were also being readied to carry Highball. Forty inert Upkeeps had been constructed for test purposes.

Two important, related tactical decisions were made. The first was that moonlight would be essential, to enable crews to drop their weapons with the necessary accuracy. Normal Bomber Command 'ops' did not take place under such conditions, which made aircraft of Harris's Main Force easy prey for night-fighters, responsible for almost three-quarters of Luftwaffe 'kills' of British bombers. Thus, the Chastise attackers would fly all the way to their targets at very low level, below the German radar threshold, where they would be hard for fighters to spot or engage. The principal menaces to the Lancasters' survival would be light flak – anti-aircraft gunfire – and such physical hazards as power lines.

Even as these issues were being thrashed out, at High Wycombe Harris made the only significant decision with which he, as a declared Chastise sceptic, was entrusted: which aircrew should fly the operation? The C-in-C determined that they should be drawn from 5 Group, an elite formation that he himself had commanded earlier in the war. He instructed its new AOC, Ralph Cochrane, that instead of diverting a line squadron to attack the dams he should form a new, special one. He also identified the officer who should lead it.

2 GIBSON

And so to Wing-Commander Guy Penrose Gibson, the man with the dog; the short, sad twenty-four-year-old with the brilliant smile who became a national hero. At the Air Ministry meeting of 15 February Saundby, newly promoted to become Harris's deputy, asserted that 'two weeks would provide sufficient time in which to train crews for this operation' – which said more about his own and his chief's insouciant attitude to Chastise than about their understanding of the supreme challenge their fliers were about to be invited to undertake. For years the vexed debate about an air attack on Germany's dams had focused upon means – devising weapons that might make possible such a stroke. Now, however, the spotlight shifted: towards the very young men, still unknowing two months before they took off, who would be called upon to fulfil the vision of Barnes Wallis.

Gibson was about to go on leave to Cornwall, after receiving a second DSO for his distinguished tenure commanding 106 Squadron, when on the afternoon of 14 March 1943 he was summoned to Grantham to see 5 Group's chilly, clever new AOC. The Hon. Ralph Cochrane, by common consent Harris's outstanding subordinate, was an autocrat possessed of better manners and more imagination than his superior at High Wycombe, especially about means of attacking Germany from the air. He was a scion of a smart Scottish family – his father was an army officer turned Unionist politician, ennobled in 1919. Ralph, one of five children, entered the Royal Naval College, Osborne, aged thirteen in 1908, the year in which the Wright brothers made their first flights in Europe.

While flying airships on convoy escort duty during the First World War, he met Barnes Wallis. Cochrane once tried to sink a German submarine by dropping four 8-lb bombs on it, without convincing either himself or the enemy of the efficacy of air power. He was still an airship man when he encountered Sir Hugh 'Boom' Trenchard, wartime commander of the Royal Flying Corps and founding father of the RAF. 'Young man,' said Trenchard, 'you're wasting your time. Go and learn to fly an aeroplane.' Within a few years, Cochrane was serving as a flight commander in Iraq, where Arthur Harris was converting

Vernon troop-carriers into bombers on his own initiative, and experimenting with the prone position for bomb-aiming. In 1937 Cochrane became a founding chief of staff for New Zealand's air force. By 1943 he was recognised as one of the RAF's ablest senior officers, described by a navigator in one of his squadrons, later the offensive's official historian, as 'a ruthless martinet'.

A bewildered Gibson was kept hanging around for several days at St Vincent's Hall, the rambling Victorian mansion boasting its own tower and spire in which 5 Group had its headquarters. He wrote: 'Group headquarters are funny places. There is an air of quiet, cold efficiency. Waafs keep running in and out with cups of tea. Tired men walk through the corridors with red files under their arms. The yellow lights over the doors of the Air Officer Commanding and [his senior staff officer] are almost always on, showing that they are engaged.' Gibson's account of his encounter with Cochrane, which appears finally to have taken place on Thursday, 18 March, is probably roughly accurate: 'in one breath he congratulated me on my bar to the DSO, in the next he suddenly said: "How would you like the idea of doing one more trip?"'

This was an extraordinary demand to make of an exhausted young officer who had already done more towards winning the war than could reasonably be asked of any man. Yet Harris did not hesitate to instruct Cochrane to make it. Whatever the C-in-C's doubts about Wallis, Upkeep and dams, he possessed sufficient guile to be determined to ensure that, since Chastise was to happen, his own brand should be stamped upon it. It was thus logical that he should nominate a protégé to lead the special squadron to be formed to fulfil Portal's fantasy. When Gibson was nominated for a second DSO, Cochrane – who appears previously to have met him only once or twice –

queried the award, suggesting that a third DFC would be more appropriate. Harris sharply overruled him: 'Any Captain who completes 172 sorties in an outstanding manner is worth two DSOs, if not a VC. Bar to DSO approved.'

Gibson wrote later about Cochrane's proposal at St Vincent's: 'I gulped. More flak, more fighters. "What kind of trip, sir?" "A pretty important one, perhaps one of the most devastating of all time. I can't tell you any more now. Do you want to do it?" I said I thought I did, trying to remember where I had left my flying kit.' Gibson momentarily supposed that he was being invited to undertake the special mission that very night, or the next. Yet for such a man as himself, who had come to know no other life save bombing, who was justly proud of being the best, the defier of fate, it was unthinkable that he should have said no.

Some commanders might have hesitated to make the request, however, in the face of Gibson's exhaustion, observed by all those who spent time with him, and manifested in severe attacks of pain in his feet. Immense labour was involved in recruiting, establishing and training a new squadron, comprising some five hundred men. In addition, its commander must meet Barnes Wallis; address technical issues with Vickers and Avro; discuss tactics with 5 Group staff officers; plan operational details; exercise in the air with whatever scratch crew he himself might assemble. Gibson did not yet know what he and his men would be asked to do, but Harris did, and Cochrane. The C-in-C and 5 Group's AOC were indeed ruthless men. How could they have fulfilled their roles, towards both the German people and their own aircrew, had they been anything else?

Three days later Cochrane saw Gibson again, and told him somewhat more: he was to form a special unit to execute the special operation, which could not take place for two months. He introduced the airman to Gp. Capt. Charles Whitworth, base commander at Scampton, four miles north of Lincoln, where 'Squadron X' would be based. Secrecy would be vital. For the time being, however, all that Gibson need know about the target was that he must train his crews and prepare his aircraft to fly very, very low. On the afternoon of Sunday, 21 March Gibson drove through the gates of Scampton bomber station to embark upon two months of extraordinary exertion, which would define his short life.

Most human beings have their demons, but the new commander of the squadron that within days became 617, the number allocated by the Air Ministry in a natural succession, was tormented by more than most. He was born in the Punjab in 1918, third child of an unhappy marriage between Leonora, a nineteen-year-old Englishwoman, and Alex, a much older officer of the Indian Forestry Service, himself born in Russia. Guy spent his early childhood pampered by the tribe of domestic attendants customary among servants of the Raj, which some of his critics claimed influenced the abruptness with which he later treated comrades of a lower caste. He was only six, however, when his mother abandoned her husband and travelled to England to place Guy in an English school.

Once 'home', Nora Gibson became an alcoholic. Her son was a fifteen-year-old pupil at St Edward's, a minor public school in Oxford, when his mother was given a three-month prison sentence for a series of offences involving dangerous driving and causing injury to pedestrians. A psychologist described her as prone to 'erratic, impulsive and excitable behaviour'. Guy thereafter saw little of her; on Christmas Eve 1939, aged forty-four, she suffered a ghastly lingering death after her clothes caught fire on becoming entangled in an electric stove while she was drunk. It is unrecorded whether Gibson heard the news immediately, but that night he became paralytically drunk at a mess party, and hurled a succession of fire extinguishers through a glass window. He makes no mention of his mother's death in his memoir, but it is impossible to doubt its impact upon him.

Ralph Cochrane once described Gibson as 'the sort of boy who would have been head prefect in any school', but at St Edward's he achieved neither distinction nor notoriety, academically or as a sportsman. Like so many of his contemporaries, the 'Lindbergh generation', he forged an early passion to fly, but his first application to join the RAF was rejected. The Battle of Britain conferred supreme glamour on fliers as the 'Brylcreem boys', jealously mocked by soldiers and sailors after an airman's image was adopted to advertise the hair cream. Before 1940, however, pilots lacked social cachet. Indeed, they had a rueful pre-war joke that a flier would sooner tell people he was a pianist in a brothel than admit to being a member of the RAF. Even after hostilities began, while many British aristocrats enlisted in the army and some in the Royal Navy, very few became pilots. The aircrew of 617 Squadron eventually included several public schoolboys and one Etonian, but none were authentic 'toffs'.

In Gibson's case, after a few months the RAF relented and accepted him for pilot training. This was indisputably exciting, but also perilous: during the inter-war years sixty-two cadets at Cranwell, the service's elite college, were killed in flying accidents. In November 1936, aged just eighteen, three months after leaving St Edward's, Gibson reported for instruction to the airfield at Yatesbury in Wiltshire. He graduated the following year, with a rating of 'average'. He ranked lower than that as a companion, however, being widely viewed, in the school-boy slang of the period, as 'bumptious'. Perhaps to compensate for a lack of physical stature, he was gauchely assertive and immodest. His determination to make a mark, to get on, was not in doubt. But his manner of setting about this, and especially his condescension to lower ranks, including ground crew, did not make him popular. He wrote later: 'I was no serviceman; I joined the Air Force in 1936 purely to learn to fly. I was due to leave the RAF [in April 1940] to become a test pilot – a good job with plenty of money in it.'

At the coming of war his squadron commander said wryly: 'Now's your chance to be a hero, Gibbo.' The young bomber pilot indeed welcomed the opportunity for advancement, as did many career warriors in all three services, but realised how slight were his chances of survival. He sought to stop his elder brother buying him a wristwatch as a present: 'Don't do it,' he said. 'I'm a dead man.'

In an unpublished draft of his memoirs, Gibson recorded a scene at the Grand Hotel in Brighton during the June 1940 Dunkirk drama, where he described himself behaving with a boorishness and casual anti-Semitism still typical of young Englishmen: 'As usual in these places the lounge was congested with doctors, businessmen and Jews who had all come out of London ... An old man who looked like a rich Jew' demanded a change of radio channel to hear the news. Gibson wrote: 'I switched off the radio quickly and the ensemble in the room suddenly became very silent. Then I swung on the old man. "Listen you stupid old bastard," I said softly, though every man and woman in the bar could hear, "do you realise that while you are here sitting on your fat behind, eating the food of the land, there are a few men fighting for your plump neck? ... I think sir you are a complete bum." And more of the same.'

Bomber Command's war made a slow start, much preoccupied with dropping leaflets over Germany, together with unsuccessful sorties against the German fleet, in one of which Gibson participated. Even when his squadron began to undertake operations over Europe, most were unsuccessful. Soon after he completed his first 'tour', he committed a folly common among young men occupying a waiting room for death: he made a precipitate marriage, in his case to an older chorus girl. He met Evelyn Moore when she was appearing in a revue, *Come Out to Play* starring Jessie Matthews, in Coventry, where he was visiting his brother, a soldier in the Warwickshires.

He was twenty-one. Eve, as she was always called, was an experienced, long-legged, somewhat coarse twenty-eight-year-old, unable to bear children: 'She was small and blonde and above all could talk,' he wrote, adding with a notable casualness: 'I was still smarting after my broken heart (so I thought) with Barbara' – whom he had characterised as 'small, dumb and blonde' – 'and as most of the boys in the Squadron claimed to have regular girls, I saw no reason why I should not have one myself and Eve looked alright to me. It was pleasant being able to live normally, going around with a nice girl for a couple of days.' Comrades speculated that their alliance, which swiftly chilled, was founded chiefly upon sexual compatibility. The wedding took place in November 1940 in Cardiff. Gibson wrote: 'This was due to Cox and King's Bank, London, who granted me an overdraft of £5 for the occasion.' Both before and after the ceremony, Gibson sustained a reputation as a compulsive sexual adventurer, who was also profoundly lonely.

By the time of his marriage he had flown thirty-seven operations in twin-engined Hampdens of 83 Squadron, receiving his first DFC. Though those 'trips' were statistically somewhat less hazardous than those undertaken by Bomber Command crews later in the war, when Germany's defences had become more formidable, all operations over Europe were desperately uncomfortable, in poorly-heated aircraft that lacked electronic navigation aids – 'beams' – such as the Luftwaffe already employed. 'Ops' were liable to end in disaster through bad weather, getting lost or both, even if the enemy did not take a hand. Gibson had seen a steady stream of comrades 'go for a Burton', their fates unknown because they simply vanished – failed to return to base.

While he was still not greatly loved by his fellow airmen, he had become keenly appreciated by his superiors. Here was a tough, effective young pilot eager not merely to do his duty, but much more than his duty, in prosecuting the war. Marginal weather never deterred Gibson from taking off, 'pressing on' to complete a mission. Yet he later wrote self-critically about this period: 'I was getting nervy, there was no doubt about it; this bombing was beginning to get me down. Sometimes over a German city we would make a half-hearted attack, knowing well in our hearts that we had not done our best.' In a passage deleted before the book's publication, but plausibly authentic, he described how a sergeant pilot once announced that he wanted to fly that night's mission to avenge his mother, who had been killed in the blitz. After briefing, a deputy flight commander carefully looked around to ensure that all senior officers had gone out of the room, then said tersely to the crews: 'Don't bother about the target tonight, chaps. A stick [of bombs] across the town will do for all, and may we kill as many people as possible to avenge Sgt. Knox's mother.'

At the time of the Luftwaffe 'blitz' on Britain there was a sudden demand for night-fighter pilots. Gibson, ever impatient to grapple the enemy, promptly volunteered, apparently with the encouragement of 5 Group's then commander, AVM

Arthur Harris, who promised him a squadron when he returned to bomber operations. He was posted to become a flight commander with 29 Squadron, flying twin-engined Beaufighters, and settled down at its Lincolnshire airfield to master the techniques of engaging German intruders with the relatively primitive airborne radar of those days. During the ensuing year he flew a hundred operational sorties – often with his new mainly-Labrador puppy as a passenger – shot down three enemy aircraft, and was awarded a second DFC. Eve stayed with him for a time, in a room above the local Lion and Royal pub, where she was understandably miserable, albeit no more so than many wartime wives. Gibson sometimes returned at lunchtime from night operations to find her huddled before the gas fire, struggling to keep warm. Yet he liked the squadron and the social scene, and wrote significantly in 1944, 'No matter what happens, I should like to record that in many ways they were the happiest [days] I have ever had.'

He survived some bad nights: once the Luftwaffe strafed the airfield as he came in to land following a sortie. He enhanced his reputation as a fiercely determined airman, but seemed frustrated that he had not achieved the big score, ace status, of some of his comrades. 'Night fighting was all very well, but it meant too much patience for me. In a year's work, involving about 70 night sorties and 20 day patrols, I had in all seen about 20 Huns. Of these I had opened fire on nine. So obviously, [bearing] in mind my score, I was not a very good shot for one thing; the other thing was that bombing was still in my blood.' Late in 1941, when Luftwaffe activity over Britain declined following the Nazi invasion of Russia, he applied to return to Bomber Command.

He was initially posted to a training unit as an instructor, normal practice to provide a break for young men wearied by the strain of operational flying. This was not remotely to Gibson's taste, however. He already displayed a passion for confronting the enemy, driven partly by ambition; possibly partly also by consciousness that his marriage was arid; partly again by a principled commitment to fulfilling the RAF's share in winning the war. Senior officers often lamented the fact that many aircrew were happy to do enough to earn their pay, while lacking fervour, emotional hostility towards the enemy. Gibson, by contrast, possessed in full measure 'Hun-hate', a quality that Harris, his former mentor at 5 Group, esteemed. As Bomber Command's new C-in-C, Harris now had authority to make good on the earlier promise to his young protégé, of a squadron command.

On 12 March 1942, as a newly-fledged twenty-three-year-old wing-commander, Gibson took over 106 Squadron at Coningsby in Lincolnshire. On his first appearance, to introduce himself to its NCOs, he administered a slashing rebuke because they failed to stand as he entered the briefing room: 'I'm afraid I was rather rude, but you have to be, so that they can know what you think.' The sergeants received badly this dressing-down from their new half-pint CO, and dubbed him 'the boy emperor'. Through the weeks and months that followed, he drove the squadron relentlessly, emphasising discipline both on the ground and in the air in a fashion that caused some to change his nickname to 'the arch-bastard'. He wrote of the two flight commanders whom he inherited, and promptly evicted: 'Neither, I am afraid, were very much good to me. One was a nervous wreck, the other was an excellent type but tired out … He was jumpy, didn't listen to what was being said. Later he got into his Flight van and went off to see his wife, who was living in Boston, and I imagined the painful scene.'

Gibson's most significant shortcoming as a leader was a lack of sympathy, or indeed mercy, for weaker vessels than himself; for aircrew who flinched before difficulties or – worse still – in the face of the enemy. Here he showed himself quite unlike Leonard Cheshire, his future successor as CO of 617 Squadron, who was supremely courageous, but also wonderfully compassionate to comrades and junior ranks.

Yet it would be mistaken to characterise Gibson as an unthinking brute, such as are many eager warriors. His book *Enemy Coast Ahead*, which was overwhelmingly his own work, is one of the best airmen's wartime memoirs, for which its author deserves a respect independent of that earned for his achievements over Germany. In one fascinating passage, Gibson speculates with some fellow fliers about the importance of the Soviet triumph at Stalingrad, while the strategic air offensive remained Britain's most significant contribution to defeating the Nazis: "'But surely you don't think bombing could win the war alone?' asked Charles [106 Squadron's adjutant, a World War I soldier]. "I don't know Charles, that is what we are taught, as you know, but we are dealing with the mass psychology of a nation [Germany] – and a bad nation at that ... They might crack, if they break out of the iron ring, or they might not. No one can say.'"

In the eyes of Harris and other messianic air-power enthusiasts, such equivocal words about the RAF's strategic air campaign amounted to heresy. Yet Gibson would repeat them publicly, emphasising his view of the limits to the potential of air power during a tour of America after the dams raid. He displayed an independence of thought, a willingness to diverge from the party line, very rare among operational pilots, or career-minded RAF officers. This did not prevent him from displaying exceptional courage and determination in the performance of his duty, but it showed that there was nothing mindless about him.

In the air, he displayed an apparent indifference to danger that terrified some of those who flew with him. In May 1942, 106 Squadron was re-equipped with the new four-engined Avro Lancaster, in which their CO often chose to fly low over Germany, on one 'trip' repeatedly circling at a thousand feet above a line of U-Boats that he spotted during a low-level return from Rostock, to gather intelligence for the targeting staff. On the ground, he embraced as like spirits 'press-on types', including three whom he later recruited to 617 Squadron. He heaped scorn on weaklings, especially those who showed themselves prone to make 'Early Returns'. To turn back once during a tour for technical reasons, without attacking the target, was a misfortune that could befall any crew. To do so more often, however, suggested accident-proneness, or something worse. One night, as a Lancaster was waiting to take off, the explosive self-destruct device on its Gee navigational aid detonated, filling the cockpit with smoke, and causing the crew hastily to abandon the aircraft on the taxiway. Gibson raced a jeep to the scene and lambasted the hapless men, whom he made plain that he suspected of 'lacking moral fibre', to use the RAF's phrase of those days. 'You've got four good engines, you'll bloody well go and bomb Germany!' he said. And so they unwillingly did.

Gibson and the later 617 pilots Dave Shannon and John Hopgood flew together on the 14 October 1942 low-level daylight raid by a formation of ninety-four Lancasters from 5 Group against the Schneider armament factory at Le Creusot in France. After bombing, Gibson and Hopgood circled their own target, the nearby power station, machine-gunning its transformers. Hopgood's aircraft returned badly holed by the detonation of its own bombs below safety height.

The CO of 106 Squadron himself seemed, in the words of Richard Morris, 'addicted to stress'. He led from the front, so that men who recoiled from his strength alongside their own privately avowed weakness, or at least vulnerability, would have been astonished to be told of his uncertainty about the war-winning potential of bombing cities. Few people ever glimpsed the chinks in his armour, the lonely youth beneath the iron exterior. One who did so was a WAAF corporal named Margaret North. Gibson's squadron had been transferred from Coningsby to Syerston in Nottinghamshire, where one night the station commander, Gus Walker, was dreadfully injured when a Lancaster's bombload exploded on the ground.

Gibson accompanied Walker in an ambulance to the nearby medical facility. While doctors worked on the group-captain, the pilot fell into conversation with a nurse from the nearby crash and burns unit. Next day he invited her out, dismissing her protests about the regulation that prohibited social inter-course between officers and other ranks. The couple began a relationship, at first platonic, which persisted intermittently through the last two years of Gibson's life. It seemed important to him – certainly more so than his flagging marriage. The couple rattled around the country lanes in his little black car, accompanied by Nigger. He talked incessantly, mostly about bombing. He also quizzed her about her work in a manner that betrayed his awareness that any night, he might himself become one of her patients.

Once Gibson walked into her unit while she was working, and sat down beside the bed of a gunner swathed in bandages who was dying of burns sustained when fire swept his aircraft after a flare exploded. This breach of his normal disdain for junior ranks was followed by a flash of warm human sympathy. He said to the young man: 'My name's Guy Gibson. Can I sit with you?' The patient was incapable of reply, but the wing-commander sat in silence beside him until he died.

In March 1943, after a year with 106 Squadron Gibson relinquished command. He had completed a total of seventy-two bomber operations, far more than most pilots survived, or were willing to endure. Unlike some squadron COs, who seldom operated, or chose relatively easy missions when they did so, he had flown some of the toughest, and made 106 one of the best-performing squadrons in 5 Group. Both humanity and common sense suggested that it was now more than time for him to quit operational flying and accept a staff job, as did most of his peers. But what else could Gibson do? His entire adult life had been devoted to bombing Germany. Stress had imposed chronic weariness. His marriage had become a pretence. He had no time or energy save for operations; no leisure pursuits save a mild interest in sailing small boats. His relationship with Margaret North seemed to depend, in considerable degree, on the fact that she, like himself and unlike Eve, was a member of the Bomber Command 'family'.

A woman spectator of one of the war's great land battles wrote of warriors who had seen much action: 'These young men, most of them not older than twenty or twenty-five in their voice and gesture, have a control and a sense of responsibility, a self-discipline, which makes them look more like paterfamilias than young men just coming from college or university.' Another woman addressed a soldier she assumed to be thirty-five, but who told her that he was twenty-three: 'The war has made me old.' She suggested that when peace came, he could retrieve his youth. He demurred: 'When the spirit is old, you can't become young again.'

Eve Gibson said in an interview long after her husband's death: 'I never really knew him. He kept his innermost thoughts to himself. His first love was the Air Force and he was married to whatever aircraft he happened to be flying at the time.' It was said of Billy Bishop, a Canadian flying ace of World War I: 'There was something about him that left one feeling that he preferred to live as he fought, in a rather hard, brittle world of his own.' Many of Guy Gibson's aircrew would have made the same assumption about their CO during those spring weeks of 1943, when not Wallis, nor Portal, nor Harris, nor Cochrane, but he himself bore the ultimate responsibility for breaching Germany's dams.

3 'A DISASTER OF THE FIRST MAGNITUDE'

The foremost challenge throughout March and April 1943 was for Wallis and Vickers to transform a metallic object, still a mere concept when Portal mandated Chastise, into a viable bomb that would bounce to the Möhne and its north-west German kin. It was unprecedented to launch aircrew training and tactical planning for an attack to be executed with a speculative weapon. Until the last week of February, Wallis envisaged his Upkeep as a near-sphere, 7'6" in diameter. However, it was discovered that such a weapon would require an intolerable delay – two years – before special steel could be made available to create the dies.

The engineer hastily reworked his design, to become instead a steel cylinder, encased in an outer 'skin' of wood, because no other workable material was available, some five feet in diameter. His initial drawings were shown to a meeting at Burhill on the evening of Sunday, 28 February. The new-model Upkeeps could be quickly cold-rolled and welded, but this change of shape was not cost-free: the ballistic properties of a sphere enabled it to run truer after release from an aircraft than would a cylinder. Later experience showed that, while some cylindrical Upkeeps sustained a straight course to the target, others displayed a leftward bias that was recognised by their inventor, and that caused them to fail to perform as intended, especially following any slight error by either pilot or bomb-aimer. The new version would contain 6,600 lb of Torpex explosive, contrasted with the 7,500 lb originally estimated to be necessary by Arthur Collins of the Road Research Laboratory. The operational bombs were filled by Royal Ordnance at Chorley, while inert dummies, loaded to an equivalent weight with a cork and cement mixture, were prepared for flight testing.

Early in March, Vickers-Armstrong reported that their Newcastle and Barrow plants could produce only around twenty Upkeeps within the necessary time-frame; their Crayford plant was manufacturing some smaller Highballs. Wallis's initial drawings were dispatched to Newcastle on the 4th, followed by more refined versions nine days later. Two further droppings of half-size dummy Upkeeps took place at Chesil in Dorset on the 8th and 9th. Craven sent Wallis an encouraging note, ironic in the light of the recent past, saying that he had met Portal and AVM Sir Wilfred Freeman – one of the assistant chiefs – twice during the previous week: 'They are taking intense interest in the whole scheme.' Yet others remained jaundiced. Saundby asked Harris on 17 March, 'Should we now send an armament officer from the Group concerned to go and obtain the necessary instruction in the use of this contrivance?' Such a characterisation of Upkeep emphasised that Saundby thought little of it. Bomber Command's deputy C-in-C, whose private passions were fishing, bird-watching and the enormous model railway layout that he kept in a room above his quarters at High Wycombe, was throughout the war no more nor less than his master's voice and ears.

Harris himself blew alternately lukewarm and cold. He now acknowledged that his chief at the Air Ministry was determined that Chastise should be executed, but continued to consider Wallis's bombs absurd. He viewed as grotesque the demands that they imposed upon the aircrew who must carry them to Germany, even by the standards of his Main Force, which often lost one aircraft in twenty committed to a given night's operations. Bomber Command possessed thirty-odd operational squadrons of heavy bombers, half of them Lancasters, on the day that its C-in-C was ordered to agree to the diversion of a unit of the latter to carry Upkeep.

These were not merely refitted standard machines, but purpose-built. Indeed, it was a bad moment for Wallis when, at meetings on 1 and 2 March, designer Roy Chadwick made plain that the engineer's airy promises about using temporarily modified aircraft to drop his bombs were unrealistic. The Type 464 'Provisioning' Lancasters must be assembled on a new production line specially created at Avro, a diversion of effort with obvious implications for deliveries of Main Force bombers. The Ministry of Aircraft Production was forecasting future output at 123 Lancasters a month, scarcely enough to replace losses and continue the build-up of Harris's Command. Even more significant, Wallis's weapon had yet to achieve a successful test drop, and informed sceptics doubted that it ever would.

On 1 April, following another meeting of AVM Norman Bottomley's Air Ministry Chastise committee, Portal's deputy reported to the chiefs of staff that construction of twenty, and not thirty, 'Provisioning' Lancasters should suffice. This was a historically important admission: study of the tactical realities made it plain that Upkeep – and, by implication, Highball also – could be employed only against undefended or lightly defended targets. It would be difficult enough to launch these weapons from bombers flying in darkness at low level, even with the advantage of surprise. Once British interest in dams was revealed, the technology of Upkeep exposed, the Germans would install defences on vulnerable targets such as would render further attacks not merely almost suicidal, but absolutely so. Low-level attackers were vulnerable to being dazzled by searchlights, shot down by flak or fatally impeded by balloon cables, all of which the Germans indeed deployed following 617 Squadron's debut. Thus, in April, it was already plain that the decisive moment for Wallis's weapons, for Bomber Command and for Germany's dams, must come at first use. There could be no progressive operational learning curve; few, if any, second chances.

A batch of twenty-three Lancaster Mark IIIs eventually emerged from the Avro factory as Type 464 aircraft to carry the weapons – twenty for Gibson's men, three for Vickers' trialling flights. In addition to removal of the bomb doors to accommodate Upkeep, a four-cylinder motor was fitted beneath the fuselage, weighing 336 pounds and manufactured by Vickers-Jassey to control submarine hydroplanes; this was to supply a hydraulic drive to the Fenner V-belt that would counter-rotate the bomb before dropping. Its speed would be adjusted by the aircraft wireless-operator, twisting a knurled wheel. Wallis was still exploring an optimum rotation rate: if the bomb was spun too fast, the charge might expand and split the casing, or even burst free of its relatively precarious retaining mechanism.

The bombs were to be held in place by hinged arms or cali-pers bolted to the aircraft's reinforced fuselage longeron. Strong springs would be fitted to force apart the calipers, releasing the rotating bomb. Wallis also had to supervise tests to establish that the spinning of a four-and-a-half-ton Upkeep would not destabilise the aircraft in flight. The additional fittings would add half a ton to the thirty-ton weight of a Lancaster. This made necessary the compensatory removal of the plane's mid-upper gun turret.

On 9 April Vickers reported that they had constructed fifty Upkeep cylinders much faster than had been thought possible; these had been shipped for filling with explosive. A gunnery range in the Thames estuary at Reculver in Kent was nomi-nated as the location for full-scale trials of the weapon, more conveniently reached from London, Weybridge and Scampton than was Chesil Beach, and with sands exposed at low tide to facilitate recovery of expended bomb fragments.

On the night of 6/7 March, Harris had launched the first of a long succession of raids that would become known as the Battle of the Ruhr, by 442 aircraft against Essen. This would continue until the autumn. One strategic principle that Bomber Command's C-in-C got right was that of concentra-tion of force. He saw that earlier attacks, scattered all over Germany with lamentable inaccuracy, failed to hit the enemy hard in any one place. The Battle of the Ruhr, some forty-four attacks on Hitler's foremost sources of steel, coal and much else between March and November 1943, inflicted significant injury on Germany's industrial base, though rather more on the residential areas of the cities that became targets.

This Ruhr campaign enabled Harris to half-reconcile himself to Chastise, because he could regard an operation against north-west Germany's dams as at least a marginal element in

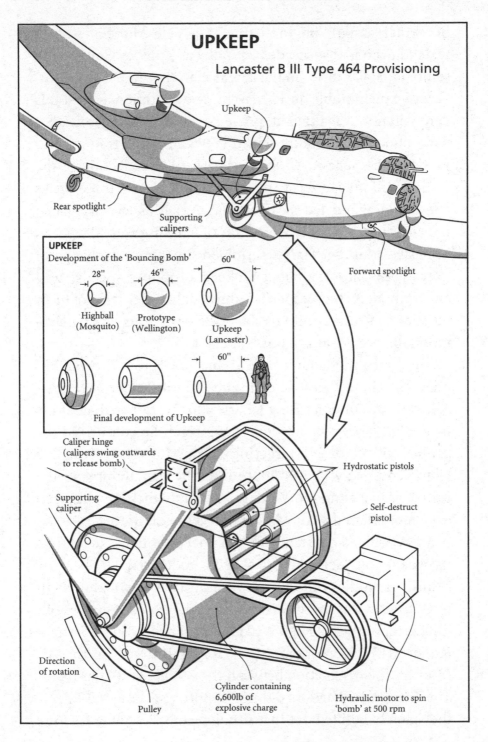

UPKEEP

Lancaster B III Type 464 Provisioning

Upkeep

Rear spotlight

Supporting calipers

Forward spotlight

UPKEEP
Development of the 'Bouncing Bomb'

28"
Highball
(Mosquito)

46"
Prototype
(Wellington)

60"
Upkeep
(Lancaster)

60"

Final development of Upkeep

Caliper hinge
(calipers swing outwards
to release bomb)

Hydrostatic pistols

Supporting
caliper

Self-destruct
pistol

Direction
of rotation

Pulley

Cylinder containing
6,600lb of
explosive charge

Hydraulic motor to spin
'bomb' at 500 rpm

his wider assault on the Ruhr. Air Vice-Marshal Robert
Oxland, who had succeeded Saundby as Bomber Command's
senior air staff officer on the latter's elevation to Deputy C-in-
C, wrote memorably on 17 March: 'The operation against this
dam will not, it is thought, prove particularly dangerous, but
will undoubtedly require skilled crews ... Some training will
no doubt be necessary.' Here was a further reflection of the
disdain still prevailing at High Wycombe, almost three weeks
after Portal mandated the operation. As late as 15 April, when
the aircrew tasked with execution of Chastise were deep into
intensive training, Harris scribbled during an exchange of
notes with Saundby about Upkeep dropping speeds, 'as I
always thought, the weapon is balmy [sic] ... Get some of these
lunatics controlled & if possible locked up.' This was barely a
month before the attack was launched.

It was a further source of grievance for Harris that the key
meetings about Chastise continued to take place at the Air
Ministry in London's King Charles Street, rather than at his
own High Wycombe headquarters, which he considered the
proper place for all targeting discussions involving his
Command. He was even more irked by the attempts of the
Ministry of Economic Warfare, a body which he abhorred
even more than the Air Ministry, to get in on the act.

MEW lobbied constantly for Bomber Command to address
its assault on Germany to specific industrial targets. The
Ministry's studies of the enemy's industrial effort were often
flawed, but at least its researchers groped for precise vulnerabilities, as the airmen at High Wycombe declined to do. In particular, an MEW memorandum of 15 March urged on the Air
Ministry the disjunction between the Möhne and Eder dams.
The Economic Warriors did not oppose attacking the Eder –
they merely pointed out that simultaneous destruction of the

Möhne and the Sorpe promised far more important results. On 27 March Portal circulated to his fellow chiefs of staff an MEW document headed 'The Economic and Moral Effects of the Destruction of the Möhne Dam and the added effects that will result from Destruction at the same time of the Sorpe and Eder Dams'. This promised that even sole destruction of the Möhne should suffice to unleash on Ruhr cities 'a disaster of the first magnitude' – a phrase borrowed from Wallis – create a serious electricity shortage, and 'have further and serious repercussions on morale'. The same was true of the Sorpe, but the Eder 'cannot be considered as a first-class objective'.

In a subsequent memorandum to the Air Ministry, on 2 April, MEW economist Oliver Lawrence showed himself much less optimistic than the chief of air staff. He thoroughly analysed north-west Germany's water and power resources, and delivered a report that was a model of its kind – far more sensible and penetrating than were most wartime industrial intelligence documents. Lawrence did not doubt that breaching the Möhne would create a local catastrophe. He was nonetheless insistently cautious about its impact on the wider Ruhr industries' water supplies. He again emphasised the irrelevance of the Eder, however vulnerable it might prove to Wallis's bombs. His central point was that while destruction of both the Möhne and the Sorpe would inflict a major disaster upon the Nazi industrial machine, the loss of one without the other would create no more than inconvenience: 'It is not possible to state that a critical shortage of water supplies in the Ruhr would be a certain and inevitable result of the destruction of the Mohne dam,' even if the physical and moral effects made the operation worthwhile. But if the Sorpe could also be breached, its destruction 'would be worth much more than twice the destruction of one'.

Yet Portal might have demanded: how many operations of war – any war – would be undertaken, if the promise of absolute, rather than relative, success was made a precondition? By April 1943 the towering realities were that the head of the Royal Air Force had committed himself to an attack on Germany's dams; that relatively large technological resources were being deployed to enable Barnes Wallis's bombs to bounce; and that the aircrew of Guy Gibson's new squadron were training day and night, at mortal risk, to implement the engineer's vision inside fifty days.

4

Men and Machines

Henry Treece, a Staffordshire writer who spent the war as an RAF intelligence officer, wrote a poem entitled 'Lincolnshire Bomber Station', of which the first verse ran:

Across the road the homesick Romans made,
The ground-mist thickens to a milky shroud;
Through flat, damp fields call sheep, mourning their dead
In cracked and timeless voices, unutterably sad,
Suffering for all the world, in Lincolnshire.

Treece could have been thinking of Scampton, through the gates of which Guy Gibson drove on the afternoon of Sunday, 21 March in a service Humber shooting-brake, accompanied by his tail-flagging, beer-drinking black friend. He had now owned Nigger for two years, and loved the dog probably more than any human being. He spoke intermittently by telephone to Eve, and she appeared briefly once or twice at the station, but their exchanges appear to have been unsatisfactory.

Warship or tank crews lived and fought amid familiar environments and a relatively even tenor of discomfort. A unique aspect of the bomber war was that Harris's aircrew spent most of their flying careers deep in the English countryside, drinking on 'stand-downs' in local pubs or dance halls such as the Gliderdrome in Boston. Then, every few nights, they were plunged into the hottest flame of battle over Germany. The revolving door between relative tranquillity and desperate peril, parallel existences in violently contrasting environments, imposed a special kind of strain. The surprise was not that some aircrew suffered more or less serious breakdowns – succumbing to what the RAF sometimes brutally branded LMF, 'Lacking Moral Fibre' – but that relatively few did so: just over a thousand men, one-third of them pilots, though a much larger number were more charitably 'taken off ops' – transferred out of operational duties to fulfil more humdrum functions.

Where infantry soldiers were merely taught to march and to kill their fellow men, aircrew had to be painstakingly instructed in the techniques of flight, at a cost of £10,000 apiece for pilots and navigators, which explains why thereafter their commanders were so unwilling to excuse them from bombing Germany. Flight training was a lengthy process, since most of it took place in safe but remote places – Australia, Canada, California, Rhodesia; some men who enlisted in 1940 did not embark upon operations until three years later. An Oxford-educated navigator described his pilgrimage from first donning air force blue to taking off for Germany as 'a course of education ... more testing and, therefore, more rewarding than any other I have received before or since. The brink of battle, too, is an interesting place.'

Scampton lay north of Lincoln, just west of the old Roman road Ermine Street – Gibson had been stationed there back in 1938. Cochrane designated it as 617's first base because half of its accommodation and facilities chanced to be empty – a unit had moved out to permit the concreting of the grass runways, leaving only one squadron operating from the station, with Harris's Main Force. Bomber bases were desolate places, as observed by Henry Treece: expanses of grass, concrete, huts and hangars occupied by a permanent population of up to fifteen hundred ground crew, administrators, parachute packers, mess staff, service police, fire teams, armourers, medics and officers' batmen, who enjoyed the privilege of relatively safe, if austere, wartime existences. They often worked hard, in all weathers, but were overwhelmingly likely to survive. Their functions were to support three hundred-odd birds of passage from two squadrons, who were likely to die.

The fliers might spend seven or eight months at a given station, but would more plausibly vanish over Germany, some night or another. In the spring of 1943, less than one man in five was completing a thirty-trip tour of operations, and only 2.5 per cent finished a second tour. No crew is recorded as having offered to Sir Arthur Harris the old Roman gladiators' farewell to the emperor, *Morituri te salutamus* – We who are about to die salute you – but that was the way things were. Statistically, wartime British and American soldiers and sailors enjoyed odds heavily weighted in favour of coming through, while in 1943 the adverse prospects of 'bomber boys' were matched only by those of submarine crews.

As a pre-war construction, Scampton was an improvement on most of the British, Canadian and American airfields that now carpeted eastern England. It was dominated by four huge hangars, laid out in a crescent facing towards the aircraft dispersals and runway. Beside these lay a hotchpotch of quarters and offices; a comfortable pre-war officers' mess; squash courts; some neat little gardens boasting daffodils, tulips, wall-flowers; mown lawns; more substantial buildings than the usual 1940s hutments. The new squadron's 'erks' – ground crew – benefited little from such indulgences, however, being packed into triple-decker bunks in what had once been the base's married quarters.

Gp. Capt. Charles Whitworth was a popular 'station-master', considerate of his men – and women. Gibson quickly comman-deered twenty-one-year-old WAAF intelligence officer Fay Gillon to handle liaison with 5 Group and details of a flying training programme, asking her only, 'Can you keep a secret?' He was pleased to discover that she was married to a soldier absent overseas, and thus presumably less likely to indulge in pillow-talk with a local boyfriend. The new squadron's officers dubbed her 'Famous'. Gibson's staff and technicians were allocated offices attached to No. 2 hangar, where he himself repaired on that Sunday afternoon to embark upon the marathon task of mustering more than five hundred men and preparing them for a very special battle. His own room at first boasted only a desk, chair and telephone, though within days he contrived at least to get it painted and carpeted, and adorned with a noticeboard and wall-maps, to which kindly Waafs sometimes added a bowl of flowers.

He made a series of swift, sound staffing decisions, sacking his first adjutant – administrative officer – in favour of the effective F/Lt. Harry Humphries, familiar to him from Syerston. To manage the ground personnel and countless office and disciplinary routines he secured from the neighbouring 57 Squadron, no doubt on Whitworth's say-so, F/Sgt. George 'Sandy' Powell, a North Welshman always addressed as 'Chiefy', and Sgt. Jim Heveron, outstanding NCOs who proved masters of organisation and improvisation.

The urgent issues facing Gibson were recruitment of pilots from among the ten squadrons of 5 Group; acquisition of aircraft; a start on low-level training. It would be weeks before the new Type 464 Lancaster IIIs were available; meanwhile, 617's crews must train on a handful of standard aircraft, diverted from other units. Bomber Command's fliers were accustomed to operating over Germany at fifteen to twenty thousand feet. When they attempted low-level manoeuvres – for instance, flying under bridges or 'beating up' airfields – they invited court-martial, even if they escaped death. Now, instead, 617's newcomers must within weeks become masters of the art. In every seven-strong crew, the man at the controls was almost as important an arbiter of their shared destiny as was luck. A Bomber Command navigator described his relegation from embryo pilot during training as 'a crushing disappointment', since the RAF was, he wrote, 'a pilots' air force'. For Chastise, the skill of the man at the controls would be trebly critical, because precision was indispensable to successful launches of Upkeep.

Sir Arthur Harris proposed manning the new squadron with tour-expired volunteer crews – in other words, veterans. His 'old lags', he asserted blithely, would not mind doing one more 'trip'. He was wrong. Recruiting suitable men posed persistent problems, never entirely resolved. Gibson assembled a core of 5 Group's best and boldest – New Zealander Les Munro, for instance, responded to a call for experienced volunteers 'for a special operation' after completing twenty-one trips with 97 Squadron. Big Les grew up on a sheep station where there was little money; he was obliged to take a correspondence course to improve his maths before becoming eligible for pilot training. He had since become respected as a stolid, determined and able flier.

Joe McCarthy, who became the squadron's sole American, also came from an uncommonly modest background for a pilot. The hefty, blond, open-faced son of a firefighter, he grew up in New York's Bronx; as a teenager he developed into a champion swimmer and occasional lifeguard, for a time on a Coney Island beach. Meanwhile on Long Island, he took some flying lessons which prompted him to try to enlist in the USAAF. Rejected because of his lack of a college degree, in May 1941 he caught a bus to Canada, where he became one of nine thousand Americans to join the RCAF.

After training and a voyage to Britain, he began operations with 97 Squadron in October 1942, and had almost completed a tour when he was telephoned personally by Guy Gibson, whom he had met and obviously favourably impressed. McCarthy explained: '[Gibson] asked me if I'd like to join a special squadron for one mission ... He couldn't tell me what we were going to do, where we were going to go, or anything ... He said, "If you can't bring the whole crew, take as many as you can. We'll probably find some for you, but we would prefer your own." I explained it to my crew and got a lot of flak back, quick. "Why? What are we going to do?" Same thing I asked, and I just had to tell them I didn't know, but it was going to be just one trip.' After a protracted crew huddle, McCarthy and his six men accepted Gibson's mysterious but flattering proposition.

To fill 617's ranks, however, its CO was obliged also to include inexperienced pilots, several of whom were summarily drafted. When, for reasons never explained, the lot fell upon Ken Brown of 44 Squadron, this dismayed country boy from Moose Jaw, Saskatchewan, said that he preferred to continue where he was – he had completed just seven operations. Impossible, Brown was told, probably untruthfully: 'It was a name transfer and we could do nothing about it.'

When the CO of 467 Squadron nominated a crew to join Gibson, in response to a circular appeal from 5 Group, their skipper took a vote. By four to three they opted to stay put, and were permitted to do so. In their place to 617 went P/O Vernon Byers, from Star City, Saskatchewan. Since the stay-behinds survived the war, while Byers and his men did not, the former may be thought to have chosen prudently. Byers' flight-school report, written in Canada little more than a year earlier, categorised him as a slow learner, but 'a dependable average pilot'. He flew his first operation to Germany in March 1943, and had completed just one more as a 'second dickey' and three with his own crew when posted to 617 – probably because somebody had to be sent, and a fickle finger pointed to him. He had no low-level experience, but nonetheless impressed Gibson sufficiently to be commissioned a few days before Chastise.

Six crews were transferred from Scampton's resident 57 Squadron, the whole of its 'C' Flight, under their commander S/Ldr. Melvin Young. Among Young's understandably bemused men, pilot Geoff Rice protested vociferously, but in vain. One of Rice's gunners, nineteen-year-old London policeman's son Tom Maynard, recorded in his diary after meeting Gibson: 'Had our interview with our new CO ... very young. He told us we were formed for a special mission of which even he hadn't the faintest idea and we have to spend the next few weeks training low level by day and by night. We changed billets to the new squadron.'

Three more crews came from 106 Squadron, and thus were well known to Gibson – F/Lt. John Hopgood; F/Sgt. Lewis Burpee, whose father managed a big department store, while his mother played in the Ottawa Symphony Orchestra; and F/Lt. David Shannon, an Australian who looked too young to have relinquished short trousers. Aged twenty, he was the son of a South Australian Assembly member. The pilot himself 'leapt at the opportunity', but among his crew only the navigator agreed to move to Scampton. The others, already familiar with life on the brink of death, baulked at a further leap into the unknown.

Gibson asked by name for F/Lt. Harold 'Micky' Martin, a slightly-built Australian whom he had met at a Buckingham Palace investiture, because he was a famous low-flying enthusiast. Martin gathered a crew of mostly Aussie mates, summoned from all over Bomber Command. Tasmanian rear-gunner Tom 'Tammy' Simpson came furthest, from a training unit at Kinloss, Scotland, where he was sulking furiously after being confined to the camp for a petty uniform violation.

Of twenty-one pilots who joined the embryo squadron in the last days of March and early April 1943, twelve were RAF, including one Australian; three RAAF; five RCAF, including Joe McCarthy; and one RNZAF, Les Munro. The RAF sustained prejudices in favour of privately-educated aircrew drawn from the British upper-middle class; the service was also wary of 'colonials': senior officers recoiled from their directness, intolerance of formality, and alleged indiscipline. A remarkable Air Ministry memorandum complained that the RAF's squadrons 'are not getting a reasonable percentage of the young men of the middle and upper classes, who are the backbone of this country, when they leave the public schools'.

In a January 1943 letter to Portal, Sir Arthur Harris likewise deplored 'the very poor type of Commanding Officer which the Dominions seem to produce. Mostly hangovers from the prehistoric past. At the best they are completely inexperienced, and at the worst they are awful.' On another occasion he vented disgust that his all-Canadian No. 6 Group boasted a venereal disease rate five times higher than that of their British counterparts. Yet less jaundiced observers, both then and since, recog-

nised that the Dominions made an outstanding contribution
to the strategic air offensive, and explicitly to Chastise, for
which they provided one-third of the aircrew. On arrival at
Scampton, George Chalmers, a bomb-aimer, thought that he
had never seen such a mix of British and Empire in one
squadron.

Although the 'white Dominions' swore allegiance to the
King, many of the young Canadians, Australians and New
Zealanders who crossed oceans to join the struggle against
Hitler at first found Britain as alien as did Americans. John
Fraser, a bomb-aimer from Vancouver Island, wrote to his
mother and sister on 15 April 1943: 'Just a year yesterday that
I landed on this foreign soil and first heard murmurs of
pounds, shillings and pence over shop counters, saw cars driving on the wrong side of the road – today those differences
plus a hundred other oddities have become quite familiar. One
can become accustomed to strange things.' Fraser was only
twenty, but engaged to a Doncaster girl named Doris
Wilkinson, whom he was aching to marry, perhaps for the
commonest reason for so many hasty marriages of that era: it
offered the surest, sometimes only, access to regular sexual
intercourse. Moreover, many young men who knew that they
were probably doomed yearned to leave something behind
them – to start a child.

By the afternoon of 27 March, 617 Squadron had acquired
some fifteen crews, who Gibson assembled in No. 2 hangar for
their first briefing. 'You're here to do a special job,' he said.
'You're here as a crack squadron, you're here to carry out a raid
on Germany which, I am told, will have startling results. Some
say it may even cut short the duration of the war. What the
target is, I can't tell you. Nor can I tell you where it is. All I can

tell you is that you will have to practise low flying all day and all night until you know how to do it with your eyes shut. If I tell you to fly to a tree in the middle of England, then I will want you to bomb that tree. If I tell you to fly through a hangar, then you will have to go through that hangar even though your wing tips might hit either side.' Discipline, he said, was essential – and so was secrecy. There would be no leave.

After the briefing broke up there was a buzz of chatter, speculation about possible targets that would persist through the seven weeks that followed. Gibson had done precious little to enlighten them; merely provided most with a first encounter with their leader, who would remain a remote figure, especially to lower ranks, together with notice that low flying would be their business. This would mean dangers extreme even by the extreme standard set by Bomber Command.

The two pilots Gibson appointed as his flight commanders were personally unknown to him, and thus must have been nominated either by Whitworth or 5 Group headquarters, probably the latter. Melvin Young, who took over 'A', was twenty-seven years old – in the context of those days, incomparably more experienced than most of his comrades. His father was a prosperous solicitor, his mother the daughter of a Californian real-estate developer. Young's childhood alternated between an American and a British life – Kent School, Connecticut, today claims him as one of its own, as does London's Westminster. In a will made a week before his death, he bequeathed £50 to each establishment.

In 1934 Young followed his father to Trinity College, Oxford, where he read law and rowed in the 1938 Boat Race for a university eight of which four were later killed flying with the RAF. He also accumulated forty flying hours with the University Air Squadron. His instructor noted: 'This member is not a natural pilot and is still rather coarse on the controls … He is very keen and has plenty of common sense.' By one of those coincidences that could have happened only in the village that was then Britain, the writer of the report was Charles Whitworth, Scampton's 1943 station commander. Young also at this time became a Freemason, and afterwards bequeathed a silver bowl to his lodge.

In 1938, while pursuing law studies in London, he was commissioned into the RAFVR, writing to a friend a few months later: 'Since we are to have a war, I am more than ever glad that I am in the air force. It is a happy, healthy life while it lasts, and I have found some old friends and made many new ones … I am not frightened of dying if that is God's will and only hope I may die doing my duty as I should. In the meantime, I remain as cheerful as most, I think, and try to keep others so.' Though graded 'above average' as both an officer and a pilot, he shared with many others a regret that he was posted to bombers rather than fighters. Even after joining a squadron, in his spare hours this big, bluff, burly oarsman continued to appear on the riverside at Oxford, either to coach a boat or to cheer on an eight. Young was unusually widely read for a wartime pilot, and Christianity was a serious force in his life: he was much dismayed when his parents separated.

During the Battle of Britain he flew a tour of operations on Whitley bombers from a base in Yorkshire which was attacked by the Luftwaffe. Fellow pilot Leonard Cheshire later recalled Young, in the airfield's shelter, producing a pack of cards as the bombs rained down and enquiring, 'Anyone like a game of bridge?' On 7 October 1940, while flying as a convoy escort – the sort of incongruous role that sometimes fell to the lot of bomber pilots in those desperate days – engine failure obliged Young to ditch in the Atlantic, from which he and his crew were retrieved by a destroyer only after a ghastly twenty-two-hour ordeal in a gale-tossed dinghy. Bill White, son of a Kansas newspaper owner, chanced to be a passenger on board the warship, and recorded the pick-up:

> One of the aviators rises wildly, unsteadily grapples at a rope,
> is too weak to wrap it around him, topples into the sea.
> Instantly a sailor goes over our rail, comes up behind the
> man with the loose-rolling head and wild eyes just out of the
> water. He ties the rope under his arms and pushes him to the
> dangling ship's ladder … Three sailors pull his sea-chilled
> body up and over the side. The others … reach the solid
> safety of steel deck, and are half-led, half-carried down to the
> cozy warmth of our wardroom.

In peacetime, after such an experience a man might require months to recover, but within weeks Young was again flying over Europe, after delivering a graphic BBC broadcast describing his 'ditching'. On 23 November 1940 his squadron participated in an attack on Turin that ended with five of its aircraft being lost through fuel exhaustion, which caused Young again to ditch in the sea, then spend ten hours tossing on a raft forty miles off Devon's Start Point before being rescued. It was after that experience that he was dubbed with the nickname with which he would die – 'Dinghy'. To survive once, in aircrew parlance, 'coming down in the drink' was notable enough; to get home twice was astonishing.

Young's godmother told a mutual American friend after seeing him following this second ordeal that he looked 'older and worn'. This was characteristic of many young men who crowded into brief wartime lives experiences and horrors – for crashing an aircraft into the ocean was indeed a horror – that should properly have been spread over decades. Young spent much of 1941 instructing on Wellingtons – no sinecure, since the accident rate at OTUs was appalling. He was then sent to a 'Wimpey' squadron which operated first out of Malta, then Egypt. In the summer of 1942 he secured a plum posting to a temporary liaison role in the United States, which enabled him to revive contact with his American connections.

Within days of his arrival, following a six-day flight, he visited old family friends named the Rawsons. Four hours after getting out of the car at their home, Ravenscroft Farm in Connecticut, he proposed marriage to their thirty-three-year-old daughter Priscilla. A tall, gawky, bespectacled girl, she was understandably amazed by his declaration, and lingered for two days before assenting. In the absolute absence of evidence, it is almost impertinent to speculate on the reasons for Young's impulsive outbreak of passion, but he emulated both Guy Gibson and Leonard Cheshire in marrying an older woman amid the stress of war. Priscilla was well-read, musical, enjoyed an independent income; but her credentials as a comfortable companion were more conspicuous than any hint of glamour.

It seems plausible that Priscilla was the only woman Young had ever enjoyed the opportunity to get to know well enough to consider as his bride, or partner in the bedroom. For all his brash mateyness, he had less self-confidence than he appeared to exude. He made plain to her that he existed on the extreme edge of peril, and would soon be subjected to this again. They were married on 10 August, just a week after his proposal, in the chapel of nearby Kent School. Priscilla told her mother, with real or assumed flippancy, that she had always fancied marrying an Englishman, 'and haven't I got a handsome one?'

Within hours of the reception at Ravenscroft Farm, the couple embarked on an eighteen-hour flight to California –

Priscilla's first experience of an aircraft – to meet Melvin's mother. The young airman spent the ensuing months among USAAF personnel at a succession of Florida, Georgia and Alabama airfields, and also gave a speech to the boys of Kent School. On 2 February 1943 he parted from Priscilla in New York, to return to operational flying in Britain: their married life had lasted just under six months.

Back with Bomber Command, Young completed a conversion course to Lancasters, then 'crewed up' and was posted to 57 Squadron. Before he had flown a single operation, and apparently at Charles Whitworth's recommendation, on 25 March his entire 'C' Flight was posted across the station. Ground crew thought Young 'officious', and complained that he came on too strongly the squadron-leader – a flight commander's rank. But Bomber Command's most effective officers were disciplinarians. Young appears to have organised much of the new unit's training programme, and to have exercised command during Gibson's frequent absences.

'B' Flight was entrusted to Henry Maudslay of 50 Squadron, who wrote with some pride and even surprise to his mother: 'My Wing-Commander here tells me they are making me a squadron-leader and giving me a Flight, but please don't take this too seriously yet – I'm not, anyway.' Born in 1921, Maudslay was the son of a modestly wealthy motor-industry executive. He grew up in rural manor houses, with holidays spent on the family's Scottish sporting estate, and at Eton became a star: Captain of Athletics, Captain of Boats, member of 'Pop'. When war came, most of his school contemporaries elected to serve in smart infantry regiments, or occasionally the Royal Navy. He, however, had cherished a passion for flight since first he flew rubber-powered balsa-wood models. In June 1940, fresh out of school, he joined the RAF, trained in Britain and then

Canada, flew a 1940–41 tour of operations on Hampdens. He was halfway through a second tour on Lancasters when summoned to Scampton and promoted.

Jack Slessor, 5 Group's able commander in 1942, contributed a word-portrait of Maudslay: 'courteous, diffident, strikingly good-looking ... an Olympian swell at Eton ... I never remember being more impressed by the quality of any young officer, and one sometimes had a horrid premonition that he was one of the dedicated sort who sooner or later would tempt providence too far.' Beyond coming from a smarter background than most of his comrades, Maudslay was also more cultured, aesthetically aware. He read middlebrow poetry, such as that of Walter de la Mare. His chosen Christmas presents for family and friends were books dispatched from Blackwell's of Oxford. If there were girls in his life, posterity knows nothing of them. Most of the letters that he wrote were addressed to his widowed mother. Mrs Maudslay visited Henry at successive airfields, and this solicitous son urged on her regular doses of Bemax, Adexolin, vitamin C and other dietary supplements.

Socially, Maudslay was a classic product of Britain's upper-middle class, with its courtesy, sense of duty, acceptance of responsibility, and perhaps also its inhibitions. He displayed insistent concern about the health problems of his old nanny, Kelly. This somewhat fastidious young man was ill at ease with the coarseness of some of his comrades, writing from a convalescent home after a motorbike smash about his relief on finding that there were no Dominions aircrew among his fellow patients, 'so the result is a definitely unstressed atmosphere ... no one is being polite, such a rest!' He continued to devour copies of *The Field* and the *Eton Chronicle*, though sobered by the latter's casualty lists. He was thrilled when, after various delays, in January 1943 he was elected a member of the RAF

Club in Piccadilly. Maudslay was unusually experienced as a flier of Lancasters, having served between operational tours with their trials and conversion unit. His surviving correspondence suggests a serious, family-minded, rurally-inclined young man, committed to fulfilment of whatever mission was entrusted to him. Of his private fears and passions we know nothing.

John Hopgood, pronounced by Gibson the best pilot on the squadron, was another twenty-one-year-old, who joined the RAF in 1940 after leaving Marlborough College and spending a few months articled to a London solicitor – his father's profession. He had hoped to go up to Cambridge, but his family's loss of its money in the 1929 Crash, and later the onset of war, intervened. Hopgood was attracted to the RAF because, as he wrote in October 1939 with an impatience widely shared among his contemporaries, 'that seems to be the only thing which actually does anything', contrasting with the army's passivity. He embraced the air force, flying one tour of bomber operations on Hampdens in 1941 and most of another on Manchesters and Lancasters in the following year. A more gregarious character than Maudslay, a fine pianist and a keen golfer, he secured the approval of Gibson as 'a good squadron type', a team player. Hopgood's most conspicuous physical feature was a prominent front tooth, about which his companions ragged him, so that he strove to avoid exposing it.

It is sometimes said that 'Hoppy' was 'Gibbo's' closest friend, but this seems speculative. In letters home the younger man enthused about one of his station commanders, Gus Walker – 'he is incredibly good and kind to us all' – and at Syerston he described a commissioned gunner named Brian Oliver as his 'best friend'. However, he never mentioned Gibson, his CO at 106 Squadron in the early months of 1943. Surviving images show a tall, lean, somewhat severe, literally tight-lipped young man, apparently committed to his demanding role. When his sister Marna became engaged to a fellow army officer, John wrote sternly to their mother: 'I'm all against wartime marriages – they are inclined to put blokes off the fighting spirit and give them additional worries and cares, which is obviously not a good thing.' Flying a heavy bomber over Germany offered a sufficiency of cares for a twenty-one-year-old from a sheltered background, who after a village church service only a few years earlier had written in praise of the sermon: 'The theme was that the greatest gift of God to mankind was the inability to see the future, and that to face the future we need God.'

'Hoppy' achieved the relatively unusual distinction of being awarded a second DFC soon after his twenty-first birthday. What is certain about his time with 617 Squadron is that Gibson trusted him implicitly as both a man and a pilot, having known him through his time at Syerston, an eternity in the context of that time and place, as he had not known Young and Maudslay. With Hopgood, as with many of his peers, it would be mistaken to be deluded by the upbeat letters to his mother into supposing that he found flying over Germany to be boyish fun. Perhaps the most vivid, moving word-portrait was drawn by his half-sister Betty, who accompanied him to an investiture at Buckingham Palace in April 1943, and saw him at home on what would prove to be his last leave: 'He looked pale and very drawn and was smoking heavily, his fingers yellow with nicotine stains. He spoke little of what he was doing (sworn to secrecy) and seemed to find comfort in playing Chopin and Mozart etc. on the piano, which he did beautifully.'

Gibson, Young, Maudslay, Hopgood and Bill Astell – an ex-Bradfield public schoolboy with a taste for foreign adventures who had spent three months at Leipzig University – represented one social strand in 617 Squadron. Many of its other pilots, however, came from humbler social backgrounds, having in common a good education and an eagerness to fly. Their supporting crew members fulfilled technical functions. More than a few who participated in Chastise had completed only a handful of operations – three flight-engineers had done none at all. This was deemed relatively unimportant. So long as they mastered their own roles, experience mattered much less among the 'tradesmen' – gunners, engineers and wireless-operators – than for pilots and navigators. For the dams attack, the proficiency and personal judgement of bomb-aimers would also assume an unusual importance.

Among wartime warriors, only tank men matched the intimacy in which a heavy bomber crew fought. A navigator called the experience of getting 'crewed-up' – customarily at an Operational Training Unit as a newly qualified flier – 'as big a headache as getting married'. They were given a fortnight 'to make a love-match', and only if they failed to create a team by spontaneous combustion were they wed by higher authority.

The strategic air offensive's future official historian and director of the Imperial War Museum Noble Frankland met an Australian bomb-aimer at a snooker table in the OTU officers' mess: 'instinctively we understood each other and we soon decided to fly together. He knew a New Zealand flight-sergeant who was a pilot [who] looked more like an orchestral conductor than a budding Bomber Command pilot … The three of us had, I believe, made the best decision of our war career.' They were joined by a Canadian sergeant rear-gunner and an English sergeant wireless-operator. 617 Squadron's Australian pilot Micky Martin went through the same process at a Heavy Conversion Unit immediately before arriving at Scampton with six men with whom he had never flown an operation, though all knew each other on the ground, and most of them shared the bond of their common country.

The RAF never satisfactorily resolved a dilemma about the award of ranks to aircrew. It struggled to impose discipline and hierarchy, in 1943 deploring some stations' practice of holding all-ranks dances. A choleric Air Ministry report of the same year asserted: 'Aircrew are becoming more and more divorced from their legitimate leaders, and their officers are forgetting, if they ever learned them, their responsibilities towards their men. Aircrew must be disabused of the idea that their sole responsibility is to fly.' These were absurdly Blimpish aspirations towards relatively highly-educated young civilians serving in air force blue only 'to do their bit' against Hitler, conscious that within weeks or months they would almost assuredly be dead.

Pilots and navigators were often commissioned, in recognition not only of superior skills, but also of the fact that many had what were deemed 'officer-like' qualities. Others, especially gunners who were not called upon to command anybody, or indeed to acquire any demanding technical qualification, were only grudgingly conceded officer status. This undisguised class chasm defied the reality of an intimacy that was inescapable, between seven men who shared in the air extraordinary perils within a single fuselage: the fate of one almost invariably proved to be the fate of all. On landing back at base after an operation, sensitive commissioned pilots found it uncomfortable to watch fellow survivors of a seven- or eight-hour ordeal disappear towards the sergeants' mess, while they themselves headed for the officers' quarters. John Hopgood wrote of his relief, at Scampton, at having three officers besides himself in his crew, 'which is much better than the all-sergeant crew which I had before'.

Micky Martin observed, from his Australian perspective, that while Hopgood, Maudslay and their kin were good pilots, 'they wouldn't have had a clue how to go out screwing with their gunners on a Saturday night'. Martin, the son of a relatively wealthy Australian doctor, was sent to England early in 1939, at the age of twenty-one, to sow some wild oats before himself pursuing medical training. This was a responsibility which he fulfilled enthusiastically, squandering a thousand pre-war pounds in eight weeks, much of it on horses – he rode in some amateur races. He had just acquired a flat in Mayfair when war intervened. After being rejected as a fighter pilot he joined Bomber Command, where he achieved a reputation as a low-flying 'press-on' type, while retaining his ground status as a hell-raiser: a fellow officer described him as 'mad as a grasshopper'. Nobody in 617 Squadron seems to have spoken or written of Martin as a friend, outside his almost all-Anzac crew – his flight-engineer was a Geordie; later in the war he was deemed unsuited to squadron command. If he was not a cerebral character, however, he was a supremely determined and skilful pilot, who really did say 'wizard prang'.

Among 150-odd others were front-gunner Vic Hill, son of a gardener at Berkeley Castle in Gloucestershire, who became David Maltby's front-gunner. Thirty-one-year-old bomb-aimer John Fort was older than most, a Yorkshireman who had served in ground crew since joining the RAF in 1932, then remustered in his new role. His counterpart Len Sumpter was a shoemaker's son who had left school at fourteen, then laboured in a steelworks before serving happily for several years in the Brigade of Guards. Pilot David Maltby was the twenty-three-year-old son of the co-headmaster and owner of a Sussex prep school, who grew up in what appears to have been a somewhat improvident middle-class environment: his father was a keen racing punter. Maltby left Marlborough College after two years, probably for financial reasons. He then spent a year as a trainee mining engineer at a Yorkshire coalpit, before joining the RAF in 1940.

F/O Harold 'Sid' Hobday was about to leave on a special navigation course when he heard that his fellow members of Les Knight's crew were moving to a special squadron. After a sleepless night's reflection, he agreed to go with them, forgoing a promotion, because loyalty to his crew seemed paramount: 'I loved the life. It may sound terrible now in peacetime to think you liked bombing people, but I liked the idea of staying as one integral part of the set-up. I wouldn't have liked the thought of another navigator taking my place in my own crew.'

Among those who would fly to the dams, there was a sprinkling of older men – eighteen were in their thirties, the oldest being a thirty-five-year-old flight-engineer. The overwhelming majority, however, were much younger. One was just eighteen, three others nineteen. Two-thirds were between twenty and twenty-five, with the most frequent age being twenty-one. The last of the first batch to arrive was Londoner Bill Ottley, twenty-year-old son of a War Office civil servant, who had joined the RAF straight from public school, but achieved a precocious reputation as a mess raconteur. He flew the last operation of his tour with 207 Squadron on 4 April, then was immediately posted to Scampton, where he arrived just two days later.

Almost all of those who now congregated variously in the officers' or sergeants' messes at Scampton were too young to have caught the movies of Fred Astaire and Ginger Rogers first time around; some were still in school when *Gone With the Wind* had its 1940 UK release. Duty, a word that invites scepticism or even derision in the twenty-first century, was a real force in the lives of many. Most smoked, with an intensity that caused a yellowing of the right-hand fingers in all save the pipe-smokers. In that age when surnames were the normal form of address between middle-class men, in or out of uniform, the mark of intimacy between aircrew comrades was not the use of first names, but instead of matey abbreviations – 'Hoppy', 'Hutch', 'Trev', or for the squadron commander – though not widely indulged – 'Gibbo' or 'Gibby'. The difference in ages between Young, Gibson and the younger men might have been only a very few years, but it represented the difference between remembering a world without war, and accepting it as their society's normal condition.

These men were destined to be branded a band of brothers, a congregation of heroes. In truth, of course, 617 comprised a mingling of the good, the bad and the ugly, such as are all gatherings of humanity. It is important not to idealise all the crew relationships. Les Munro described his Canadian gunners as 'rather hard cases', and some of his comrades characterised each other in harsher terms. Socially, many of these men, who served together for only a matter of weeks, had little in common. The most that can be said is that a crew which had forged a bond through extended operational experience together, such as that of Australian Les Knight, shared a mutual confidence lacked by those of their brethren who had come together more lately. Almost all of the generation that fought for the Allies in World War II enjoyed a far more coherent view of what they fought against – the huge evil of Nazism – than of what they were fighting *for*, which no politician had bothered credibly to tell them.

* * *

By early April 1943, 617 Squadron comprised over five hundred aircrew and ground personnel – fifty-eight officers and 481 NCOs. A reflective British airman of the previous war, Cecil Lewis, only nineteen when he flew SE5 fighters on the Western Front, wrote: 'We lived supremely for the moment. Our pre-occupation was the next patrol, our horizon the next leave. Sometimes, jokingly, as one discusses winning the Derby Sweep, we would plan our lives "after the War". But this had no substantial significance. It was a dream, conjecturable as heaven, resembling no life we knew. We were trained with one object – to kill. We had one hope – to live.'

Most of this remained valid for Lewis's lineal descendants a generation later, the men of 617 Squadron, except that few thought of themselves as killers. It has been a peculiarity of warfare in the past century that the young men carrying out aerial bombardments have been granted a moral absolution, partly because of the risks they themselves have borne. Partly also, they were physically remote from their victims: unlike many soldiers, they were not obliged to look into the eyes of the people they killed. Guy Gibson's men were overwhelmingly preoccupied with their own duties and vulnerability as they prepared to carry out a mission which, they were promised, could be of incalculable value to the Allied cause, and to the liberation of Europe.

Eileen Strawson, a local farmer's daughter enrolled in the WAAF, became Gibson's driver. One day as they approached Lincoln, they collided with a cockerel which was trapped flapping in the front fender of their Dodge. After the girl unhesitatingly wrung its neck, Gibson said squeamishly, 'I don't know how you could do it.' She shrugged: 'I've killed hundreds of chickens.' John Hopgood's elder sister Marna asserted after his death: 'One of the reasons he went into Bomber Command was that he didn't wish to see the immediate results of human suffering from the weapons of war. He felt that to be in the clouds would separate him from the awful act of killing. I know this worried him a great deal.' It seems a significant irony that during those weeks in which Gibson and his men prepared themselves for the supreme ordeal of Chastise, the last thing on their minds was the notion of becoming instruments of death for anyone save, most plausibly, themselves.

2 FLYING

The squadron was officially formed on 23 March, and next day adjutant Harry Humphries reported its first recruits ready to start flying, though they were handicapped by lack of aircraft. The specially modified Upkeep Lancasters would not be ready for weeks; meanwhile they made a start with standard versions borrowed for training, though there were not enough even of these for each crew to have its own.

Of the three British heavy bombers of the war, the Avro Lancaster was the undisputed star. Seven thousand were ultimately built, against six thousand Handley-Page Halifaxes and two thousand Short Stirlings. Pilots have a saying about aircraft: 'If it looks right, it is right.' The 'Lanc' looked superb, especially in the air, inspiring an affection among its crews unmatched by any rival. Each example, powered by four Rolls-Royce Merlin engines mostly then built by the US Packard company, cost the British taxpayer £42,000. Cruising at 216 mph and carrying the bombload of two American Flying Fortresses at a ceiling of twenty thousand feet, the Lancaster ranks with the Spitfire, Mosquito, Fortress, Mustang and Dakota among the great Allied aircraft design successes of the Second World War.

Beneath, behind and before its great glasshouse of a cockpit canopy, five crew members clustered almost within a handshake of each other. The bomb-aimer manned the twin front-guns until they neared the target, when he adopted a prone position, peering down through his sight at the cauldron of a German city, amidst which illuminant markers glowed, positioned with variable accuracy by Pathfinders. For Chastise, of course, no such guidance would be available: each aircraft's bomb-aimer would be called upon to determine the destiny of his own Upkeep. Immediately behind and above the nose positions the pilot sat in his big armoured seat, clasping the control yoke. Beside him, on a little folding seat, sat the flight-engineer, who in 1942 had replaced the second pilot, to conserve on the huge expense of pilot training – and the wastage of losses. The engineer, almost always an NCO, monitored instruments and handled throttles under the pilot's supervision. Gibson's John Pulford, a former 'erk' like all the Chastise engineers, had been a motor mechanic in civilian life; his counterpart in Melvin Young's crew, Dave Horsfall, was a graduate of the RAF's Halton apprentices' school.

Further back on the port side of the cockpit, facing outward, the navigator occupied a little curtained cubby-hole, poring over his maps beneath a pinpoint light and monitoring the screen of the Gee electronic navigation aid, which provided reliable assistance until the plane crossed the Dutch coast but only intermittently thereafter, because of German jamming. Navigators were important people, whose lapses could kill their comrades almost as readily as those of pilots. David Maltby's Vivian Nicholson was a twenty-year-old former join-er's apprentice from Newcastle, for whom Chastise would be his first operation; Gibson's Harlo 'Terry' Taerum, a Canadian, had been an academically outstanding schoolboy and also a fine sportsman.

Further back still, facing forward on the port side, sat the wireless-operator, often the most bored man in the aircraft, because he had little to do save sustain a listening watch through his headphones, or peer out of a tiny window. He was also the hottest crew member, because the heater-inlet from the aircraft's port inner engine was located immediately below his seat. Most crews conducted a running argument about the cockpit temperature, influenced by whoever sat furthest from the warmth. Bob Hutchison, Gibson's wireless-operator, was a teetotaller with a girlfriend named Twink in nearby Boston. Bill Townsend's w/op F/Sgt. George Chalmers had chosen his pilot because he explicitly wanted to fly with an all-NCO crew: 'You had more comradeship.' Maltby's Antony Stone was a barber's son from Winchester, his father a Russian Jewish immigrant who had married a Hampshire girl after army service in World War I.

Aft of this cluster of humanity loomed the thigh-high bulk of the main spar, which it was necessary to clamber over to reach the rear of the fuselage. This was crammed with equipment: the Distant-Reading compass; a squat box containing the Gee receiving equipment, linked to a screen in the navigator's cubby-hole; ammunition cans and runways carrying belts to the turrets; a rest-bed for a wounded man; the Elsan portable toilet, which aircrew used warily – in freezing conditions, a man might leave the skin of his bottom stuck to the seat. Behind all this clutter was the loneliest place in Bomber Command – the aircraft's rear turret, in which only a small man might feel remotely comfortable, and which only a brave one cared to occupy at all: there was no room for a rear-gunner to wear his parachute. If the plane was hit, unless he jumped fast – and some 'hit the silk' without waiting to check whether his pilot might keep flying – he was unlikely to get out at all.

Gibson's new rear-gunner, also appointed squadron Gunnery Leader, was an unusual figure in holding the rank of flight-lieutenant and owning a small estate in Wales. Richard Trevor-Roper, universally known as 'Trev', was older than most at twenty-seven, and had flown over fifty operations. His earlier career had been rumbustious, as remarked by his kinsman Hugh, historian and wartime MI6 officer, who wrote in June 1942: 'I've never met my cousin Dick of Plas Teg, but whenever his name drifts into my ken it is attached to some exploit showing a proper spirit of enterprise and adventure – either controlling an extensive underground betting organisation as a schoolboy at Wellington, or speedtrack-racing, or climbing the outside of skyscrapers … or disgracing the name of Trevor-Roper by being cashiered from the Regular Army, or rehabilitating it by brilliant exploits in the RAF. I expect he's an awful bounder really, but among my drab and dreary relatives … he beacons from afar and helps me to bear the burden of my name. I look with relief from them to him, as one lonely mountain might eye another across a flat and featureless waste.'

'Trev', an unusually large man for a rear-gunner, became famous, or notorious, as 617's foremost hell-raiser on the ground, his excesses earning the respect even of Australians and Canadians who found some of the squadron's other officers tiresomely 'tight-arsed'. He was unusual also in being a year married, his wife one of four wedded to members of the squadron who were then pregnant. More than a few people thought Trevor-Roper a bully. His studied rudeness irked Humphries, the adjutant, who described him as resembling 'a very sinister stage villain'. F/O Jack Buckley, Dave Shannon's rear-gunner, ranked close to Martin and Trevor-Roper among the serious drinkers, but was also a fine golfer and an ace clay-pigeon shot.

A navigator observed that while on operations most aircraft were dominated by their pilots, his own evolved a different system: 'When we were under attack, the gunners became the captains, and when we were in the target area the bomb-aimer was the captain. When we were not under attack, not taking off or landing, and not in the target area, I was the captain. This was an unspoken system which we evolved spontaneously, and it produced in our crew the nearest possible approximation to a single brain aided by seven pairs of hands.'

Chastise prompted changes both in the accustomed lay-out of 617's aircraft, and in the responsibilities of crew members. With mid-upper turrets removed from the Type 464 Lancasters to save weight, their occupants were instead committed to man the front guns, which at low level would be granted unusual importance. Conventional bomb-sights became redundant. Other new responsibilities evolved during the training and experimentation which Gibson orchestrated. On 27 March he received from 5 Group a directive that told him enough to make a start: 'No. 617 Squadron will be required to attack a number of lightly-defended special targets … These attacks will necessitate low level navigation over enemy territory in moonlight with a final approach to the target at 100ft at a precise speed, which will be about 240 mph.'

The key element of early training, Gibson was told, would be to practise very accurate navigation in moonlight or simulated moonlight – this was a phase of the war at which it was not unknown for Bomber Command's Main Force to attack the wrong German city. For 617 Squadron, a deck-level approach should escape the attentions of enemy radar and fighters, at the price of rendering Gee almost useless. Gibson was also instructed to have his crews practise bomb runs over water, in which pilots and bomb-aimers must attain a mean accuracy of no more than forty yards.

5 Group identified nine lakes and reservoirs in Wales and the North Midlands as suitable practice targets. One of the new arrivals, Bill Astell, who had joined from Scampton's neighbouring 57 Squadron, was immediately dispatched into the air with orders to photograph all the areas designated by Grantham. Four days later, some fliers started navigational practice on routes which encompassed the listed reservoirs, though two crews – those of Joe McCarthy and Bill Townsend – overcame a general ban on leave to secure four days' absence, partly because they lacked aircraft to fly; partly because McCarthy's bomb-aimer, George 'Johnny' Johnson, had for weeks been scheduled to celebrate his wedding to a Torquay girl, Gwyn Morgan, on 3 April. His forceful American skipper insisted that the marriage should take place before Johnson did any more bombing.

Gibson had his first meeting with Barnes Wallis at Burhill on the afternoon of 24 March. Mutt Summers collected him from the local railway station in a little Fiat, and drove him to the Vickers design offices. Gibson later described the encounter, in a passage of his memoir which has become famous.

> Wallis said: 'I'm glad you've come; I don't suppose you know what for.'
> 'No idea, I'm afraid. [5 Group] said you would tell me nearly everything – whatever that means.'
> He raised his eyebrows. 'Do you mean to say you don't know the target?' he asked.
> 'Not the faintest idea.'
> 'That makes it awkward – very awkward.'

Wallis dared not breach security to enlighten him. Instead, Gibson had to be content with being shown the films of trial drops of half-size dummy Upkeeps, and given an explanation of the general principles of the weapon that his men would unleash.

On 29 March Gibson drove thirty miles south to Grantham, where Cochrane belatedly admitted him to the secret of 617's targets, for his own ears only. The pilot's first sensation was relief that he and his squadron were not to be called upon to fulfil his nightmare of attacking *Tirpitz* in its Norwegian fjord refuge. The German battleship was defended by countless anti-aircraft guns: vain attempts to sink it had already cost the lives of scores of RAF and Royal Navy aircrew. Thus enlightened, and once more accompanied by Mutt Summers, Gibson went back to Wallis. The engineer explained details of the dams' construction, together with their importance to German industry. It fell to the pilot to disabuse his host of one crazy delusion: Wallis enquired whether it might be feasible to mount the attack in daylight. Gibson responded emphatically that the only possibility of success lay through attacking under moonlight. Then he returned to Scampton, to direct the low-flying programme.

Some days later, probably on 4 April, 617's CO flew a Lancaster to the Derwent reservoir near Sheffield, to gain a cockpit-view of a low-level attack over water, with Young and Hopgood as his passengers. Despite haze and surrounding hills, at mid-afternoon the task seemed easy enough. When, however, they repeated their approach again and again, as the early spring light faded it became progressively tougher. At dusk they suffered terrifying seconds, when their aircraft seemed doomed to plunge into the reservoir.

* * *

During those days, and through the weeks that followed, Gibson was obliged to live a hurricane schedule of handling 'bumph' – his least favourite task – while conducting briefings before exercises and post-mortems afterwards; making his own practice flights with a crew whom he needed to get to know; dashing intermittently to Grantham, the Air Ministry, and the bombing range at Reculver. His responsibility was made no easier to bear because he was unable to share with any comrade the secret of the targets. He leaned heavily on the pilots whom he knew best and thus trusted most, especially Hopgood and Shannon. Everything was done by example. A bomb-aimer recalled: 'He said, "If I can't do it, you can't do it. But if I can do it, then you can do it."'

He forged one important new relationship – with Barnes Wallis, despite or perhaps because of the fact that the fifty-five-year-old engineer was easily old enough to be his father. Gibson was seized by the soaring imagination of Wallis's conception. Wallis, in his turn, was moved by the assurance, competence and authority of this young man, a warrior such as he had always felt nagged by guilt for having himself failed to become in the earlier war. Wallis the Victorian, already a teenager when the old Queen died, was now working in intimate partnership, mutual dependence, with a techno-warrior who was supremely a product of the mid-twentieth century, the new age.

Throughout April, crews flew intensively, often in threes, whenever weather permitted and there were serviceable aircraft for them to use. The theory behind the formations, in which Gibson's own force eventually flew to the dams, was that gunners could provide mutual protection in the event of fighter attack. As for navigation, before reaching the targets on their big night, they would need to steer a succession of doglegs across Holland and Germany, for which rigorous adherence to plotted courses and pinpoint turns would be essential, to escape flak concentrations. If the Lancasters strayed into gunners' sights, the enemy's 37mm and 20mm automatic weapons could engage them at almost point-blank range.

'We'd work between turning-points on legs of about fifty or sixty miles,' said bomb-aimer Len Sumpter. 'You'd fly fifteen minutes this way, and fifteen minutes that way.' Turning-points were selected for their resemblance – undisclosed to crews – to features they would encounter over Germany: a bend in the Trent did duty for a similar curve in the Rhine. A bomb-aimer or navigator would periodically warn his pilot – for instance – 'Thrapston just coming up about two miles to port, skipper. Go on five degrees to port.' Vernon Byers' crew carried out one exercise that took them over Morayshire, to the delight of their Scots flight-engineer Alastair Taylor, who wrote to his mother: 'I hope we didn't scare you too much last Monday. I saw you and Aunt Julia just in front of the house but I could not pick dad out anywhere, so thought he would probably be at a pig sale.' This would prove Mrs Taylor's last fleeting glimpse of her son.

In order to simulate night-flight in daylight, the squadron secured access to yellow-tinted goggles and to American equipment for overlaying cockpit canopies and bomb-aimers' blisters with blue celluloid. There was only enough tinted plastic to fit three aircraft, but most crews secured a turn in its use. Initially, each 'blued-out' Lancaster carried a 'safety pilot', without goggles, as a lookout in case of emergencies. Dave Shannon said that when using the equipment 'one had a strong urge to tear off the goggles and see what the hell was happening'. They overcame this, however. Most found it thrilling, suddenly to be empowered to hurtle over rural England in a fashion hitherto rigorously denied to them, stampeding cattle, sheep, sometimes pedestrians and motorists. Flight-engineer Ray Grayston said: 'It's like riding a motor-bike at 100 mph.' Nigger was a frequent passenger on Gibson's flights, and Waaf Fay Gillon an occasional one, either with the CO or Micky Martin.

The British people had grown accustomed to living surrounded by the toys of war, both on the ground and in the sky. However, thirty-ton Lancasters zooming over their heads, so low that flashes of the crews' pink features were discernible beneath their brown leather helmets, seemed unreasonably alarming. Some civilians, especially farmers, assumed that the strafers were sensation-seekers defying regulations, and filed complaints to 5 Group's harassed staff. Wireless-operator George Chalmers said: 'There was one particular farm and … round about lunchtime we used to go down to the hundred-foot level. Right on our track was this haystack that these Land Girls were building … We came over the farmhouse and a couple of horses leapt over the gate and dashed across the field.'

Wireless-operator Jack Guterman, an unusually cultured young airman, wrote to his sister on 20 April while he was airborne: 'Another "tour" of Cornwall on such a glorious afternoon is not to be missed … At the moment we are flying low over a bottle-green sea; blue cloud shadows thrown from a clear sky scattered with slow-moving lozenge clouds, make dark patches in the water and a long strip of dazzling gold denotes the sandy shore of the English coastline. What could be more idyllic? Proust would have made something of it.' Yet some fliers discovered that extreme low flying promoted air sickness, which the medical officer dispensed pills to assuage.

While each member of Bomber Command belonged to a Group and a squadron, in battle in the night sky over Germany

– and each operation over Germany was a battle – fellow occupants of an aircraft shared the only loyalty that mattered.
Before coming to Scampton, Guterman was due to be posted
to a training unit after completing his tour of operations.
Instead, however, he chose to follow to 617 his pilot Bill Ottley,
with whom he had formed a friendship founded upon their
shared enthusiasm for literature and art. Outside their own
crews, most men referred to each other as 'bods', and preferred
not to get close. 'You said "Good morning" to them,' said Len
Sumpter, 'but you never really got intimate. You were your
own little band of seven – you kept yourselves to yourselves …
You'd got to keep your brain on the job. You couldn't afford to
relax or think of anything else.'

Early Chastise training tested to the limit each aircraft's
navigator. Les Munro's Jock Rumbles, a Scottish schoolmaster's
son who had been washed out of pilot training, was rated
outstanding. Others, however, were found wanting. Gibson
demanded the removal of the navigator of a pilot named
George Lancaster, who had been transferred from 57 Squadron.
This caused the rest of the crew to close ranks. One goes and
we all go, they said, and insisted upon being returned to their
old unit. Another crew from 57, that of F/Sgt. Ray Lovell, were
found 'not to come up to the standard necessary', and were
likewise relegated. These frictions showed up the error that
had probably been made in transferring Melvin Young's flight
wholesale to the new 'special op' squadron. Several other individuals were posted out of 617 for unspecified reasons, at least
one because he found its activities frankly terrifying.

Lest it should be supposed that Gibson's emphasis on navigation was exaggerated, several of those crews later lost on
Chastise fell victim to imprecise wind monitoring and
map-reading, which caused them to drift into the sights of

enemy gunners. Conversely, those in the navigators' seats who got it right on the night deserve more applause than they sometimes receive. Each crew refined its own methods of map-reading at low level, some by using the bomb-aimer to unroll and monitor a long custom-cut strip map of their route as they flew, using water features and railways as the most reliable markers. Identifying pinpoints as they passed proved almost impossible. Instead, Les Munro's bomb-aimer Jim Clay, a thirty-two-year-old Tynesider, wrote: 'The trick was 1) to keep your map(s) orientated 2) pick out salient features ahead, or to either side 3) pass the pinpoint to the pilot and navigator 4) mark your map and check back.' From the outset it was obvious that waterways – lakes, rivers and canals – would offer the most readily identifiable navigational markers.

In their off-duty hours, most men stayed on the station. Dinghy Young liked to sit cross-legged on the floor of the officers' mess, holding his beer by the body of the tankard rather than the handle, and helping himself to an occasional pinch of snuff rather than the usual cigarettes or tobacco. Some fliers were less riotous than others. Les Knight, much respected as a pilot, was among a significant number who took religion seriously. A practising Methodist, the quiet, serious, slightly-built Australian neither drank nor smoked, never dated a girl, and was unable to drive a car or even ride a bike. Henry Maudslay, the former Eton Captain of Boats, a gentleman at all points, was a less hearty drinking companion than Dave Shannon or Micky Martin – and how could it have been otherwise? It was all a longish way, albeit a shortish time, from the days when Maudslay, clad in plus-fours, was shooting grouse on the family estate beside Loch Shin.

Others loved to party. A pilot described a typical drunken mess-night stunt: 'The "kneeling behind trick" was performed until the floor crunched with glass, faces blacked with cork, coloured with lipstick. Those from outside, who had never seen anything like it before, looked on amazed that human beings could act in such a barbaric way.' This was Guy Gibson, writing of a riot in faraway times, before he assumed the huge responsibility for leading the dams raid. At Scampton both he and Nigger were happy to share an evening pint with his pilots, but his glass-crunching days were over.

In the mess halls where they ate, the banter between aircrew and girls of the WAAF staff – by no means all of whom would have been deemed objects of desire in civilian life – was relentlessly sexual, in a way that is hard to imagine seven decades later. 'What's the collective noun for Waafs?' 'A mattress!' A Downham Market WAAF officer named Bedworth was invariably addressed by her male comrades as Bedworthy. One of the women said: 'We took it all in good part because we knew the great strain they were under and the dangers they would soon face.' Moreover, for all the crude jokes, many of the young men who made them or laughed at them went to their graves as innocent as the day they were born. There was only a minority of married or heavy-dating aircrew, who dashed to meet loved ones whenever they were released. David Maltby had a wife, Nina, in lodgings at Woodhall Spa. Richard Trevor-Roper's pregnant wife occupied similar quarters in coastal Skegness, while Gibson's Eve stayed in London. There was precious little glamour in the life of a 'bomber boy's other half.

Almost all the men, even the relatively ill-educated, inhabited a world in which handwritten correspondence, both incoming and outgoing, assumed immense importance. Telephone contact with families was difficult and expensive, brief and unsatisfying, even when connections could be made. The fliers' letters, where these survive, form their principal testament: they had few opportunities to build and sustain lasting relationships with girls. The sentiments which they expressed were as immature as might be expected from recent schoolboys.

Richard Hillary, author of *The Last Enemy*, observed a few months before his own death in January 1943 that an intelligent and cultured person who met aircrew on the ground might be moved to disdain by their 'lack of awareness, perception and sensibility … the dull pilot with a pint of beer in one hand and the *Daily Mirror* in the other – that pilot whose emotional reaction to the most cultivated woman or the local barmaid is identical – to him they are simply women'. But Hillary argued that a scornful witness would miss the critical point, 'that that boy's life lies up in the sky'; that a flier acquired an intense sensibility 'through the combination of great mobility and great power, alone with wind and stars'. Few aircrew matched Hillary's lofty intellectual vision, but he was surely right, that what even the most callow of his peers experienced in the sky made them more interesting human beings than their student frolicking on the ground might suggest.

What makes a good pilot, and especially a pilot capable of meeting the extraordinary demands imposed upon 617 Squadron in 1943? Most airmen answer such a question in phrases that include the word discipline, together with sensitivity to the moods and attitudes of their aircraft. The very act of flying a heavier-than-air machine requires a defiance of nature, an exposure to risk. Guy Gibson was viewed as a good, rather than a great, handler of a Lancaster. This immensely highly-strung young man nonetheless possessed exceptional commitment, concentration, strength of will. Micky Martin, who was indeed a 'natural', prided himself on the fact that, while he and his crew were notorious hell-raisers on the ground, in the air they never ceased to take pains. The pilot moved his head constantly, to avoid being distracted or deluded by optical illusions beyond the windscreen, or indeed by blemishes upon it. On Main Force operations, Martin tried never to fly a steady course, but instead ducked and weaved, banking slightly to examine the darkness below and on both sides of the aircraft. Every man in a good crew save the navigator was always watching, watching, watching everything beyond the Perspex of cockpit and turrets.

A pilot said about airmanship: 'You need always to be afraid of the danger of flying, to remember that at any moment an aircraft can jump up and bite you.' Although virtuoso trainees tended to be posted to fighters, there was no *genus* fighter or bomber pilot: instead, the job made the man. Flying a Lancaster required considerable physical strength, especially in the sort of manoeuvres demanded by Chastise. Many wartime pilots, even after two years' training, found that the mere responsibilities for taking off, flying and landing a four-engined aircraft made oppressive demands upon them even before the enemy intervened. No bomber was designed to be nimble – to respond instantly to slight movements of the controls – as was a fighter. The Lancaster's role was that of a load-carrier, not an acrobat. Its many protrusions defied the principles of streamlining, aerodynamics; dragged down its speed. This made it all the more astonishing, that 617's pilots achieved what they did, and unsurprising that some were unable to meet the special demands made by Upkeep.

As the pressure of training increased, uncertainty and anticipation mounted, tempers frayed. There were tensions between Dinghy Young and his newly-chosen crew: he expected to be addressed in the air as 'sir' rather than the more familiar 'skip-

Three great dams of north-west Germany: the Möhne (top) – in 1943 the roofs of its towers were removed to emplace flak guns; the Eder (above), vulnerable to Barnes Wallis's bombs, but irrelevant to the Ruhr water supply; the Sorpe (bottom), vital to Nazi industry, but a poor target for 'bouncing bombs'.

Barnes Wallis, then 37, with his
beloved Molly, 20, in the year
of their marriage, 1925.

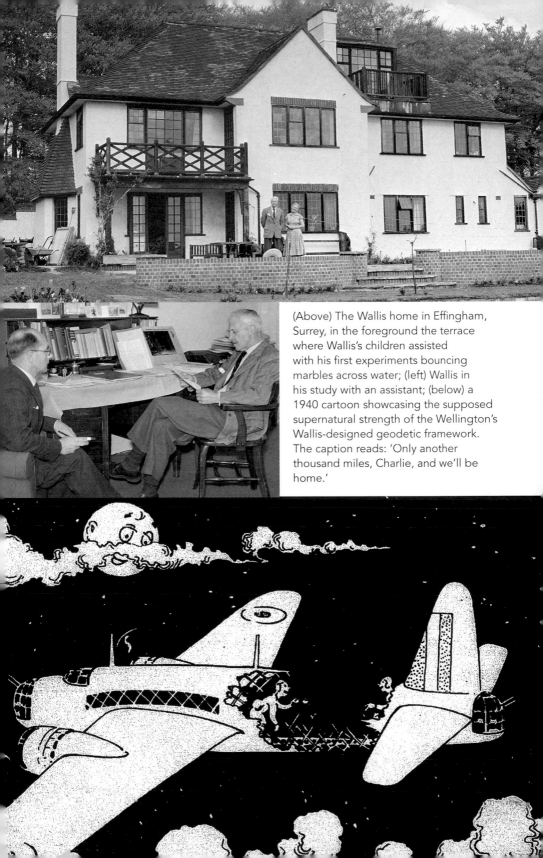

(Above) The Wallis home in Effingham, Surrey, in the foreground the terrace where Wallis's children assisted with his first experiments bouncing marbles across water; (left) Wallis in his study with an assistant; (below) a 1940 cartoon showcasing the supposed supernatural strength of the Wellington's Wallis-designed geodetic framework. The caption reads: 'Only another thousand miles, Charlie, and we'll be home.'

Testbeds for the vital experiments which demonstrated the feasibility of destroying masonry dams with an explosive charge small enough to be carried by the RAF's heaviest bomber, the Avro Lancaster. (Above) Nant-y-gro dam eight minutes after the first, unsuccessful test explosion and (below) after it was breached. (Left) the 1/50th scale model of the Möhne constructed at Watford.

(Above) British warlords: Cherwell, Portal, Pound, Churchill. (Below) Critical figures in the Chastise saga: (left to right) Cochrane, Craven, Bottomley, Collins, Linnell, Winterbotham.

(Left) Air Chief Marshal Sir Arthur Harris, the formidable, obsessive 1942–45 C-in-C of Bomber Command, an implacable foe of Chastise though he belatedly seized public credit for it. (Below) Harris in an unaccustomedly mellow mood at Springfields, his wartime official residence, with his second wife Jill, 27 to his 50, and their three-year-old daughter Jackie.

One of the very few images of Gibson piloting his modified Lancaster G-George, here seen over Reculver; (below) an Upkeep bounces enthusiastically onto the seafront watched by Wallis (left in the picture) and RAF officers including Gibson.

A 617 Squadron Lancaster drops a test Upkeep at Reculver, and suffers serious damage from dropping too close to the sea. This is believed to have been Henry Maudslay's aircraft, which consequently proved unserviceable for the attack.

per'. Some Dominion aircrew, especially, were irritated by his booming voice and Oxford accent, which contrasted with the quiet, understated manner of Henry Maudslay. Front-gunner Gordon Yeo, a twenty-one year-old Welshman who had been washed out of pilot training in Canada, wrote to his parents about Young, his own pilot, in frankly grudging terms a few days before Chastise: 'He is not so bad lately, I expect that is because we are getting used to him, but he is the cause more or less for us not getting leave.'

Gibson was a harsh disciplinarian to ground and aircrew alike. F/Sgt. Ken Brown thought the CO ran the squadron as if he was head prefect of a school – when the Canadian pilot appeared late at a briefing, he was ordered to wash all the windows of the flight offices. Gibson could also be boorish, as he showed himself towards a young airman named Turner, a 'runner' who performed odd ground jobs around the squadron. One day Gibson shouted 'George!' towards Turner. The boy continued on his way until the wing-commander grabbed him as he entered the orderly room. 'You bloody fool!' bellowed Gibson. 'When I call you, it's an order. I've a good mind to put you on a charge. Take this envelope to "A" Flight commander.'

Shortly afterwards, Turner called on the adjutant. 'Excuse me, sir. I'm a bit upset. The Wing-Commander just rousted me.' Humphries said, 'Well, why didn't you stop when he called you?' The airman responded, 'My name isn't George, sir ... I thought the CO was calling someone else.' Humphries reported the young man's dismay to Gibson, who promptly summoned him, then banged the table. 'If I call you George and you have no further intelligence than to think I am talking to a brick wall, then I am sorry for you. Now, in future you are George. You are George, understand, George?' Turner muttered, 'Oh, um, yes sir,' and retired.

This footling story shows how, amid the stress of fulfilling his extraordinary responsibility, Gibson sometimes reverted to crassness, especially with junior ranks. His driver, Eileen Strawson, said: 'He wasn't very well liked, I'm afraid.' She thought that she herself got on well with her boss because, being much older, she mothered him. Meanwhile Dave Shannon, who had flown with Gibson on 106 Squadron, thought him 'one of the finest leaders of men that I've ever met'. No reasonable person, conscious of the circumstances of 617's CO, would indict him for his lapses. But they explain why, as he prepared his men for the most testing ordeal of their lives, he was the focus of much respect, little affection.

5

The Brink of Battle

1 SIXTY FEET

The squadron's April bombing practices took place at the RAF's Wainfleet range on the Wash, twenty minutes' flying time east of Scampton, where two huge white wooden cricket sightscreens were erected to approximate – unknown to the crews – the sluice towers on the Möhne and Eder dams, respectively 639 and 781 feet apart. During the last week of April, thirty-one such exercises were carried out, in which 284 bombs were dropped, at an average distance from the markers of thirty-one yards. The projectiles were not Upkeeps nor even dummy Upkeeps; instead they were 11½-pound practice bombs, and most crews were still obliged to fly production Lancasters, rather than the Type 464 variants they would take to Germany when their operation was launched. The first modified aircraft reached Scampton on 8 April; by the end of the month there were just twelve, of which nine were serviceable. Neither fliers nor erks much liked the look of these 'abortions', with twin 'clappers' – the calipers for embracing Upkeeps – protruding clumsily beneath the doorless bomb bay.

The pilots steadily gained experience and confidence in low-level flight, at the cost of some terrifying moments which

caused planes to return to the airfield with branches and fol-
iage wedged in their underparts and tail wheel. Gibson's own
bomb-aimer 'Spam' Spafford exploded over the intercom in
broad Australian after one low pass, 'Christ! This is bloody
dangerous!' According to Gibson, he meant it. Joe McCarthy,
flying one day at a hundred feet, was enraged – because
appalled – to be passed underneath by another Lancaster, later
admitted to be that of Les Munro.

All the planes borrowed for training purposes were battered
by the stresses to which they were subjected: one came home
with its electronic IFF – Identification Friend and Foe – unit
torn from its mounting by a practice bomb that bounced up
and hit the fuselage. On another badly-buffeted aircraft four
bolts, which held the outer panel to the centre section,
suddenly sheared. Flight-engineer Ray Grayston said: 'It's
amazing to me that the Lancaster[s] stayed in one piece, the
way we had to fly [them].' On 9 April Norm Barlow's aircraft
suffered a bird-strike which caused it to graze treetops, badly
damaging two engines and smashing the Perspex of the
bomb-aimer's and flight-engineer's canopies. Yet when the
pilot wrote home to his mother in the Melbourne suburbs,
like most of his comrades he contrived to portray his daily
routine as a lark: 'I am now at a new squadron that is just
forming, hence we shall not be operating for some weeks, you
will be pleased to know, all we do is fly, fly, fly, getting plenty
of training in. Today I flew for five hours with two other crews
doing low level formation flying it was really good fun … I
have a practically new crew now, you can hardly blame the
boys for wanting a rest after all the trips we have done over
there [to Germany].'

Dinghy Young landed exhausted after the second night's
bombing practice. He told Gibson, 'It's no use. I can't see how

we're going to do it.' And that was while they were still practising attacks launched at an altitude of 150 feet. For all their
immaturity, they thoroughly understood that a moment's carelessness would be fatal in the manoeuvres which they were
practising to execute over Germany. Len Sumpter said: 'It was
when you'd got to a peak and you couldn't get any better as a
crew, that was when you most had to keep up your concentration. Every time you had to do the flight as though it was your
first or second ... Otherwise, if the pilot didn't concentrate
you'd probably hit a tree or pylon.'

By the time Chastise was launched, David Maltby, for
instance, had flown twenty-three training flights in six weeks,
totalling twenty-nine hours airborne, and had dropped forty-
one practice bombs, with an average error of forty-one yards.
At the end of April's intensive training, Joe McCarthy had
accumulated almost thirty-seven hours' low-level daylight
experience, twelve and a half night flying; David Shannon had
flown twenty-seven daylight, eleven night hours. Gibson, by
contrast, was so overwhelmed by other responsibilities that he
had been able to fly less than twenty daylight hours, just four
and a half in darkness.

Two improvisations, intended to address key problems
posed by Chastise, emphasised the range of imaginations that
contributed to the operation. The first was for bomb-aiming.
At extreme low level no conventional bomb-sight, nor any
contemporary electronic device, could fix the moment at
which an attacking aircraft should release its Upkeep, 450
yards from a dam wall. W/Cdr. Charles Dann, supervisor of
aeronautics at Boscombe Down experimental station,
proposed a palliative. The two towers on the Möhne were
known to be 639 feet apart. Following some calculations, he
created a wooden triangle, with a peephole at its base through

which the aimer peered at twin markers – nails positioned some nine inches ahead and apart. The bomb-release toggle should be pressed at the instant the markers aligned with the Möhne's towers. The nails could be moved further apart before approaching the Eder, in recognition of the greater distance between its towers.

Gibson tested the primitive Dann sight with its creator on an approach to the towers of the Derwent dam. It seemed to work. Yet the experiment took place in the absence of the vibration imposed on a 'Provisioning' Lancaster's airframe by the rotation of Upkeep, and without the distractions of enemy fire and the racket from a front-gunner blazing his Brownings overhead. A bomb-aimer could spare only one hand to steady the sight, while the other clutched the bomb-release toggle. Most crews indeed used Dann's creation on the night, but some – for instance Dave Shannon and his bomb-aimer Len Sumpter – made private choices to substitute chinagraph marks on the Perspex of the nose blister, with strings drawn back taut to the bridge of the bomb-aimer's nose. Neither the Dann sight nor the alternative home-made aid offered a reliable bomb-aimer's tool: both were improvised expedients, not solutions. To succeed, Wallis's bomb required absolute precision: the limitations of the Dann sight may explain the imperfect releases of most Upkeeps on the night of Chastise. Personal judgement and luck would play decisive roles in the outcome of each aircraft's attack.

The second challenge was to establish, and to maintain, exact height above the water. Ben Lockspeiser, director of scientific research at the Ministry of Aircraft Production and a veteran member of the dam research 'family', solved this one entirely successfully. He recalled research on the potential of spotlights to guide Coastal Command aircraft attacking submarines at night. These had failed above choppy open seas, but Lockspeiser persuaded Cochrane that on a calm inland lake, the technique would work. After testing various positions for fitting the spotlights to a 'Provisioning' Lancaster, one was set behind the bomb bay, another in the forward camera recess. If the navigator – who had no other immediate responsibility during the attack – peered down through the starboard Perspex blister protruding from the cockpit, he could direct the pilot to lose or gain height until the beams of the Aldis lamps formed a figure of eight, two touching circles.

Maudslay flew the first aircraft to Farnborough to have the lamps fitted, returning to meet a sceptical reception from Scampton's aircrew, which was echoed in a witheringly scornful note penned by Sir Arthur Harris when he heard of the technique. The C-in-C ridiculed the notion that 'his' crews could or should fly over Germany showing spotlights. The system was first tested in darkness above the runway, then over water. The doubters were confounded, and Gibson became satisfied that the experiments proved its workability, although into May technicians were grappling with such problems as leaking oil clouding the lens of the rear Aldis behind the bomb bay. Because no member of 617 Squadron was killed training for Chastise, the scale of risk and the intensity of stress inherent in each one of those April flights are sometimes missed. Dave Shannon said, 'One thought that the chap calling "Down, down, down" was never going to stop.'

And even as 617 Squadron laboured in the air, Wallis and the Vickers team were still struggling to make Upkeep viable. On Sunday, 11 April, a modified Wellington and a dummy under-sized bomb, 4ft.6in. in diameter, were pronounced ready for a full trial, following balancing and spinning tests. Next day, the

12th, Guy Gibson and his bombing leader Bob Hay travelled two hundred miles to Kent to see their prospective weapon in action. At Manston airfield, however, they were told that Barnes Wallis and his team were still fine-tuning the equipment: there would be no trial that day. The squadron commander experienced the almost unprecedented sensation of finding himself at a loose end. He and Hay drove through Margate; the boarded-up, desolate seaside resort was still wired against invaders, and the only available treats were fish and chips. Having eaten these, they idled away a few hours in the sunshine beside the sea.

That night Wallis wrote to his adored Molly from a local boarding house, where he had with difficulty found a bed: 'My own sweetheart Darling, Just a brief line to tell that I am alive and well.' He concluded, 'Ever your adoring husband, Barnes.' On the back of the envelope dispatched to Effingham, made from an old Tate & Lyle sugar packet, this droll, witty man wrote:

Oh Censor! Should thy steely Eye
Upon this missive chance to pry,
Then prithee for the Writer weep,
His feet are COLD; he cannot SLEEP.
Spokeshave-on-Spur

Next morning, 13 April, Gibson and Hay reported to the ruined St Mary's church, which was serving as base for the trials, inside the perimeter of the test range. Two Highballs had already been dropped from a Mosquito. At 0920, as they stood watching on Reculver's shingled beachfront together with Wallis, AVM Linnell, Ben Lockspeiser and two camera teams, a twin-engined Wellington flying at 289 mph dropped a half-sized dummy Upkeep, spinning at 520 rpm, from a height of eighty feet. It fell away from the aircraft to smash into the sea with such force that the surrounding woodwork – the 'outer skin' – immediately shattered. Wallis took heart, indeed was hugely encouraged: the inner steel cylinder bounced onward, unscathed.

Yet a second experiment at 1108, with a dummy Upkeep released from a Lancaster, its rotation speed reduced to 300 rpm and the height increased to 250 feet, caused the bomb's outer casing once again to break up. Wallis said, 'Oh, my God.' A deeply sympathetic Gibson watched: 'The foaming water settled down again, and there was a long silence ... I thought he was going to have a fit ... I can remember clearly the sight of that cold, quiet figure standing sometimes alone, sometimes with two or three others, on the brink of the water, looking up at the great Lancaster. There was a tenseness about the way in which he stood, with his legs apart and his chin thrust out and a fearful expectancy about everything.'

Wallis waded into the sea to recover fragments of the bomb, and ordered immediate work by the Vickers engineering team to reinforce the next dummy. Gibson wrote: 'In our secret hangar [at Manston] that afternoon there was feverish activity as the mine was strengthened ... Small men with glasses toiled side by side with sweating airmen, none of whom knew the use for which the thing was intended. They had no meals, and worked through an [air-raid] alert without pause.'

Gibson and Hay passed the intervening hours by borrowing a single-engined Miles Magister, in which they explored the bombing-range from the air. This jaunt almost ended disastrously: their engine suddenly cut out, and every nearby field was laced with anti-invasion obstacles. They wrecked the plane in the ensuing crash-landing, and were fortunate to climb out unscathed. A local man hastened over, shaking his head and providing them with their only laugh of the day when he said, 'I think they teach you young fellows to fly too early.' A policeman who insisted on taking a statement said he was glad to see that Kent's anti-aircraft-landing devices worked so well.

The two airmen returned to Scampton without waiting to see Upkeep's third trial that evening. The drop, from a height of a mere fifty feet, again caused the wooden casing to shatter, though the steel cylinder rolled on. On 15 April a Bomber Command staff officer reported to Sir Arthur Harris: 'No further trials are to take place until next Friday at the earliest, when it is hoped that from an examination of all the available data some modification will be possible to prevent the mine breaking up on impact. The general impression I got was pessimistic.' When Cochrane sought urgent guidance from Ben Lockspeiser's team about whether Upkeep was likely to be ready by the due date for the attack, barely a month hence, he was told: 'It is impossible to make any definite statement at the moment.'

On 17 April a crowd of distinguished and influential spectators once more assembled at Reculver to witness a renewal of tests. At 1245, the scheduled hour for a drop, visibility was too poor for flying. Wallis and others assuaged their impatience by stripping naked and plunging into the sea, to derisive comments from the shore. Next day, Sunday the 18th, Mutt Summers took off from nearby Manston, and dropped three successive spheres. Two sank, while the wooden casing of the third once again broke up. Wallis's imagination was seized, however, hope rekindled, by a sudden revelation: once again the steel cylinder had sailed onwards. Why bother with the outer 'skin'? He ordered all the wood to be stripped from the remaining dummy bombs at Manston.

Yet on 21 April, when flight testing was resumed in Wallis's absence, a cylinder sank immediately after being released at low level. Next day another steel dummy, released from 185 feet at 260 mph, shattered and vanished beneath the waves. Upkeep's begetter recognised that these trials, less than a month before Chastise was scheduled to take place, were 'complete failures'. But he also believed that he saw a way through. He had become convinced that Upkeep would work if released from a lower altitude, and thus at a flatter angle, from an aircraft flying more slowly: collision with the water would thus impose a lesser shock upon the bomb. He specified a speed of 232 mph and a height of just sixty feet – on the ground, a distance of twenty yards.

Was this feasible? Could the pilots of 617 Squadron perform such an extraordinary feat of airmanship, in the face of the enemy? On 24 April, Guy Gibson was flown to Weybridge from Scampton for a meeting with Wallis, who was visibly exhausted. The engineer said: 'The whole thing is going to be a failure unless we can jiggle around with our heights and speed.'

'What do you mean?'

Wallis advanced his extraordinary proposition. Could pilots attack at an altitude of sixty feet – fly heavy bombers straight and level at night over water, under conditions which ensured that a split-second lapse would mean death? It is hard to over-state the magnitude of the burden Wallis thus thrust upon Gibson. He was inviting 617 Squadron's CO to embark on an undertaking far beyond the accustomed requirements of war, and beyond even those of the bomber offensive. Purpose-built torpedo-bombers customarily attacked at a hundred feet over the open sea. In any normal circumstances an older, more mature man, mindful of scores of lives other than his own at stake, might have declined. A less ruthless figure than Wallis – and this gentle man indeed displayed absolute ruthlessness in pursuit of his purposes – would have felt unable to make such a request of Gibson and his comrades, which in the event would cost the lives of at least one crew, perhaps two. But this was war; a conflict during which even Western commanders, who professed to recoil from Japan's kamikaze suicide warriors, not infrequently ordered their own young men to undertake missions that would almost assuredly result in their extinction. For all the expressed concern of Harris for 'his' crews' welfare, he and his subordinates persisted for three years in dispatching them on operations – routine Main Force 'trips' – that they knew would terminate the lives of a majority, before they completed their 'tours'.

Gibson never flinched from making brutal demands on those under his command – more than a few of whom, in consequence, disliked and feared him. He repeatedly displayed a contempt for the odds against himself. Here was the pilot who, two months earlier, had made three successive bomb runs over Berlin to ensure accuracy, with BBC reporter Richard Dimbleby as his passenger. That performance reflected not merely courage, but also what his schoolmasters, a few years earlier, might have denounced as 'showing off'.

But 'showoffs' are of priceless value to wartime commanders. And to boffins. If Gibson had told Wallis on 24 April that what he wanted could not reasonably be asked from the pilots of his squadron, most of whom had negligible low-flying experience, Chastise would probably never have taken place. A cluster of vultures, some of them clad in the heavily brass-bound service caps of air marshals, were impatient to declare Wallis's project dead. As it was, however, Gibson returned to Scampton to tell his men that they must train to launch an attack lower than a heavy-bomber squadron had ever attempted by day, never mind by night. And the engineer resumed his intense, desperate efforts to create a workable version of Upkeep, a bare three weeks before the dams were to be attacked.

At Scampton, tensions had developed between the aircrew of 617 Squadron, engaged on apparently incessant training duties, and others who were operating almost nightly over Germany. A popular impression persists that the base was dedicated solely to the launching of Chastise, yet 57 Squadron was also flying Lancasters from its grass runways, and suffering its share of Main Force's losses. On the night of the 16th, Bomber Command staged disastrously costly twin raids on Pilsen and Mannheim, from which fifty-four aircraft failed to return. Charlie Williams, Norm Barlow's fellow-Australian wireless-operator, wrote: 'What price the last big raid. The losses were rather staggering. But I think at least they must have pranged the place well and truly.' They had not. The Pilsen bombs fell upon an asylum for the insane, instead of the Skoda works.

On 28 April a further series of dropping trials began at Reculver, initially with dummy Highballs, which the Royal Navy was still chafing to employ. At 0915 on the 29th, Gibson was again a spectator as a full-size steel cylindrical Upkeep, rotating at 510 rpm, was released from a Lancaster flying at an airspeed of 258 mph, at an altitude of sixty feet. The bomb bounced, bounced, bounced – six times in all – travelling 670 yards across the sea before sinking. Its only flaw was a leftward bias, which was to prove inherent in the design. In another trial next day – a series of London meetings intervened, attended by both airmen and engineers – a bomb released at sixty-five feet and a speed of 218 mph ran for 435 yards and bounced four times. On May Day, Gibson rang Wallis from Scampton to say that he was confident that, assuming the Upkeeps were filled and delivered in time, his squadron could execute Operation Chastise. A further test at Reculver next morning, from a height of eighty feet, was less successful, but Wallis was not much alarmed; he attributed the failure solely to a rough sea.

Thus at the eleventh hour, or rather fifteen days before the German dams were attacked, he had his triumph. The inventor, together with the technicians of Vickers and Avro, was satisfied. The weapon, in the new cylindrical form fixed only following the late-April tests, could achieve the effect for which he had been groping for a year. Subsequent Reculver trialling, which continued through the first fortnight of May, was directed towards determining optimum aircraft speed, together with bomb revolutions.

Meanwhile Vickers personnel conducted parallel experiments to ensure the stability, balance and general airworthiness of Lancasters fitted with Upkeeps and their ancillary equipment, which had vital implications for those who must fly them to Germany. The final width of the bomb had to be reduced by just over an inch, from an original sixty-one inches, to fit beneath the bay of the Type 464 Lancaster. As late as 6 May, Avro engineers were still experimenting on a test aircraft to fix specifications for the caliper arms that would hold the bombs. To enable Chastise to happen, scores of hands and brains continued to make their contributions until the last moment before 617 Squadron took off for Germany.

The three-year timeline between the first British declarations of enthusiasm for attacking dams in 1940, and the approach to maturity of the bouncing bombs in May 1943, may seem grotesquely protracted. It would have been impossible, however, for industry or the Royal Air Force to find means to bring the project to fulfilment at any earlier period of the war. Wallis's hour was now at hand, because his remarkable conception achieved a rendezvous with time, place, the Lancaster bomber and Gibson's pilots such as had been beyond Britain's means in the desperate days. The launching of Chastise reflected the turning of the tide; it represented a small but symbolic step towards passage of the initiative in the historic struggle against Nazism to the Allied camp.

2 'NO NEWS THAT WOULD INTEREST YOU FROM HERE'

On 4 May, in Gibson's weekly report to the station commander, he declared his crews 'ready to operate'. He said that his gunners proposed to use 100 per cent daylight tracer ammunition in their Brownings, instead of the usual one round in four, as a 'frightener' against German searchlight and flak crews. This was almost certainly a bad decision, which did more to pinpoint the Lancasters than to unnerve the defenders on the ground. Tracer also burned out barrels very fast. But British bomber crews valued the moral effect of firing back at Germans who were shooting at them, and 617 Squadron's were no exceptions.

Security remained a nagging concern. 'Tammy' Simpson, Micky Martin's rear-gunner, a student lawyer from Tasmania, recorded in his diary on 22 April: 'Bags of security gen and various threats to those who are found from now on speaking or writing about our job … I think it is something to do with combatting U-Boats.' Gibson lambasted one officer in front of a large group of aircrew for indiscreet use of the telephone. All calls out of Scampton were being monitored, and one of Gibson's techniques of command was to deliver rebukes before an audience, thus making them humiliations. A navigator said: 'He made you feel … about two inches high.'

The pilots of reconnaissance aircraft photographing the target reservoirs were ordered to bale out in any emergency over enemy territory so that their Spitfires would crash, destroying cameras and films in order to prevent the Germans from discovering that the pilots had been scrutinising dams. Within hours of awakening to such a threat, the defenders could adopt measures that would transform Chastise from a supremely hazardous operation into a suicidal one. Those in the secret of the targets pored over each day's photographs, studying water levels. By 17 April this was fifteen feet from the top of the Möhne dam wall. At the beginning of May it was only ten feet, and still rising.

On 2 May, Gibson dispatched an intemperate handwritten note to 5 Group, expressing fury that the squadron's armament officer, Henry 'Doc' Watson, 'knows more about the operation than either of my flight commanders and at that time, more than I did myself'. This was not Watson's fault, but that of other officers at Manston, where he had spent the previous month alongside Vickers personnel, observing procedures for handling 'Provisioning' Lancasters and Upkeeps. On Watson's return, he reported to the CO that staff officers had told him about the objectives, which seemed a shocking breach of security. Gibson swore Watson to secrecy before firing off his angry protest to St Vincent's. This concluded, with the formality characteristic of official correspondence in that era: 'I have the honour to be, Sir, your obedient servant Guy Gibson W/C.' As the CO laboured amid so many pressures, his fears were understandable.

Meanwhile, Cochrane entertained to a generous lunch the boss of the Stewarts & Lloyds steelworks at Corby in Northamptonshire, to secure permission to use as a training site its nearby water source, recorded on Chastise documents as 'Uppingham Lake'. The big white boards on the Wainfleet range were dismantled on 3 May, carted west to the Eyebrook reservoir, two miles south of Uppingham, and reassembled atop its dam. Throughout the fortnight that followed, this became the focus of 617's training by day and night. Local people, initially alarmed, grew accustomed to the shadowy forms of Lancasters roaring overhead, firing red Very flares as they pulled out after bomb-dropping. Little knots of civilian spectators gathered nearby to marvel at the performances. Some practices also took place over Abberton reservoir near Colchester, which had nothing in common with Germany's Eder, but lay at a convenient distance from Scampton.

Cochrane visited the airfield on 30 April, accompanied by a flurry of 'bull and brass' as unfamiliar to airmen as it was unpopular. He was followed on 5 May by old Lord Trenchard, still proud to be identified as the father of strategic bombing, as well as of the RAF, of which he had become first chieftain on 1 April 1918. The next day Bomber Command's C-in-C, who rarely left High Wycombe, astonished Whitworth and his officers by making an appearance. It does not seem overly cynical to suggest that, while Sir Arthur Harris still privately deplored Chastise, he knew that it had become a focus of intense excitement among Britain's directors of war. He was thus determined that any success should be identified with himself, though he had said and written enough to be able to distance himself, should it prove a fiasco.

At a 5 May meeting at the Air Ministry, chaired by Norman Bottomley, there was discussion of water heights in the German reservoirs. Given the technical issues still outstanding, should the attack be postponed a month? Absolutely not, said Wallis. For Upkeep to work as designed, levels needed to be five feet or less below the crests of the dam walls, and as spring advanced they would fall ten feet per month. He had now determined that crews should be instructed to drop their bombs from an altitude of sixty feet, flying at an airspeed of 210–220 mph, with bomb-aimers pressing release-toggles at a distance of 410 yards from the dam walls. The period 14–26 May was agreed as the window for launching Chastise, when crews could also expect the moonlight indispensable for seeing their targets.

A few days earlier, Gibson had recognised the exhaustion of his men. Dave Shannon said, '[The training] seemed an age at that particular time.' The CO relaxed his ironclad resistance to allowing leave: they were all granted two days' absence, with a few men getting three. Canadian bomb-aimer John Fraser, aged twenty, seized the opportunity to marry his beloved Doris, aged nineteen. The newlyweds were thus able to spend one night together before he flew to the dams. Meanwhile, Gibson knew that he himself was worn out: 'I began to get ill, too, and irritable and bad-tempered, and of all things, there began to grow on my face a large carbuncle.' The medical officer told the squadron commander with ironic humour that the only cure would be a fortnight's leave. Gibson laughed, and took away a bottle of tonic. His WAAF driver remarked upon one manifestation of the strain that he suffered in those days: more than once, as she drove him to conferences, he asked her to pull off the road. He would climb out and gaze for a few moments into the sky, lost in thought. Then they drove on.

His only private interlude was a discreet meeting, the first in three months, with WAAF nurse Margaret North, now married. He asked her, 'Are you happy?'

'I suppose so. How's Eve?'

'All right. How's Douglas?'

'All right.'

The symmetry of their stagnant emotional lives caused both to burst into spontaneous laughter.

On 6 May Gibson presided over a conference of his pilots. The operation for which they had been training, he said, now seemed likely to be launched within a fortnight. Meanwhile, he planned a dress rehearsal. Three formations of three aircraft – the crews that had proved most successful in training both as navigators and bombers – would fly designated night routes to the Eyebrook reservoir, then attack singly at precisely sixty feet and 232 mph, as directed by Gibson overhead. Here was another important innovation, in which Chastise pioneered a pattern for the future of the British bomber war. By the 9th all 617 aircraft were fitted with VHF voice-radio communication, enabling Gibson to exercise personal command of his crews over the target for the first time in the offensive against Germany. Amid so much else to learn, sets were installed in crewrooms, so that pilots and wireless-operators could familiarise themselves with procedures, and practise channel-switching.

In the dress rehearsal, after attacking the Eyebrook, crews would fly on to the Abberton reservoir and repeat the process. Meanwhile six other crews would conduct approach runs against the Derwent dam, simulating the lateral approach that would be used at the Sorpe. The remainder, forming a flexible reserve force on the night, to be directed while they were airborne to whichever target 5 Group nominated, would meanwhile conduct a bombing practice at Wainfleet. Watson, the armaments officer, was told that all the Torpex-filled Upkeeps must be ready by 12 May.

Wallis spent the morning of the 6th working on fuel tanks for the Windsor bomber, before driving to Reculver for further Upkeep tests that afternoon and the next. Thereafter, for three days bad weather interrupted flying. On 9 May S/Ldr. 'Fab' Fawssett, a Bomber Command intelligence officer, dispatched to 5 Group the latest brief on the German dam defences. The Möhne alone possessed flak: the sloping roofs on its towers had been removed to make way for platforms, each holding a 20mm gun; there was a third such weapon on the wall beyond the northern tower, and three more in emplacements below the dam. A feeble attempt had lately been made to camouflage the entire structure by placing along it a line of fake fir trees; it seemed possible that there was also a searchlight position. Some heavier 88mm guns, deployed on the Möhne since early in the war, had been relocated a few months earlier to defend targets under immediate RAF attack. Meanwhile the Eder and Sorpe seemed unprotected.

On that same 10 May, 5 Group dispatched to Whitworth at Scampton a handwritten draft operation order for Chastise, which he was instructed to discuss with Gibson, then return within two days offering proposed amendments. This order closely followed the attack pattern decreed for the dress rehearsal against the Eyebrook and Abberton. A 'first team' of nine aircraft would seek to breach the Möhne, then make for the Eder. Five more would attack the Sorpe, without benefit of VHF control or any airborne commander, while the other six aircraft – a total of twenty were expected at that time to take off – formed the mobile reserve. It was an extraordinarily ambitious concept, that a squadron should be directed and redirected over German targets in darkness, during a period of two to three hours, wherein the defenders might be expected to become galvanised. The operation order, No. B976, anticipated that 'destruction of the dam may take some time to become apparent and careful reconnaissance may be necessary to distinguish between breaching of the dam and the spilling over the top which will follow each explosion'.

Gibson made one important change of plan: 5 Group proposed that the attackers should climb to three thousand feet to cross the enemy coast, heavily defended by light flak, then descend again to low level. He instead stipulated low level all the way, except for occasional upward surges to five hundred feet to check landmarks. He was probably correct. There was no good choice for bombers entering German-defended airspace: at three thousand feet, enemy radar would assuredly pick them up, fighters would be scrambled. Another issue for the planners was that it was desirable to stagger the arrival of aircraft at designated objectives. It seems facetious to speak of congestion over the dams, but there would be a real risk of collision if too many Lancasters circled in a constricted airspace. They would thus take off in succession, with the Sorpe wave scheduled to go first, and the squadron's several elements routed to cross the enemy coast at two different points, a hundred miles apart. Gibson wisely vetoed a 5 Group suggestion, that Mosquitoes should stage an exploratory 'nuisance raid' on Soest, ten miles north of the Möhne, an hour before 617 arrived. Such a mission was far more likely to wake up local Germans than to secure useful last-minute intelligence, or disrupt the defences.

What proved Henry Maudslay's last letter home, dispatched on 10 May, was full of domestic detail: 'My dear Mummy ... Mrs. Gibbons, out of the blue, has sent me the most lovely morocco leather wallet ... I can think of no news that would interest you from here – except that I will probably send some socks home for darning! ... It has been infernally cold again, and I hope that you are not suffering from it too much. Perhaps Mrs. Fisher makes a good treacle pudding! Best love to you all from Henry.' News that would greatly have interested and terrified Mrs Maudslay, but which he was unable to share with her, included her son's recent return from a low-flying exercise with tree fragments trapped in his aircraft.

Next day, 11 May, the first live Upkeeps were delivered to Scampton, still warm to the touch after being filled with Torpex, to be inspected with bewilderment and near disbelief by their prospective carriers. Les Knight's bomb-aimer Edward Johnson said: 'It just didn't seem possible that this enormous piece of machinery could revolve … The idea of this thing whizzing around in the aircraft and then being released just didn't seem feasible.' That day, Gibson, Hopgood and Martin dropped practice bombs at Reculver, the Australian inflicting some damage on his aircraft. On the following evening, the 12th, Henry Maudslay and his crew suffered a nightmare experience as they took their turn with an inert Upkeep. In daylight, lacking the aid of spotlights to hold their height steady, they slid too low on the approach over the sea. As their bomb smashed into the waves, a rising mass of water struck the speeding Lancaster 'like a gigantic fist', inflicting serious damage on its tail, lower fuselage and elevators, and drenching the rear-gunner, carter's son Norman Burrows, fortunately a resilient young man from Liverpool's tough Toxteth district.

The plane flew on, trailing streams of water from its bomb bay and rear turret. Pilot and crew were spared from disaster by the extreme good fortune that none of the four engines was affected. They limped higher, some of the crew surely shaken, and landed safely at Scampton. The aircraft was rushed into a hangar for repairs. In those days, ground fitters and mechanics were worked to exhaustion, sometimes denied meal breaks, chronically short of sleep as they strove to achieve maximum serviceability for the 'Provisioning' Lancasters in advance of the operation. Gibson wrote of 'that empty noise which all hangars have ... the ringing of metal against metal as cowlings were being beaten in and dents knocked out ... the raucous voice of some fitter singing his weary love-song'. An immediate Court of Inquiry about the Reculver mishap found that the pilot had misjudged his height and was flying below fifty feet; but under the circumstances no further action was thought appropriate. Nonetheless, it is probably significant that Gibson nominated Hopgood and Young, respectively, to assume command at the Möhne and the Eder if he himself went down: he considered them likely to make a better job than Maudslay of 'running the show'.

Les Munro had an experience almost as alarming as Maudslay during the same sequence of daylight practice drops. Fractionally misjudging his height in the absence of help from the spotlights, he 'attacked' too low. The bouncing bomb hit the aircraft, jamming Canadian rear-gunner Harvey Weeks in his turret. Three more 617 aircraft dropped Upkeeps at Reculver – those of Shannon, Knight and Barlow. Afterwards Shannon's bomb-aimer, Len Sumpter, had his first and only personal encounter with his squadron commander before the operation, receiving a mild rebuke for dropping his bomb twenty yards short of the aiming mark. Absolute concentration was essential, said the CO, and Sumpter agreed: 'When you see so many things in front of you through your sight, your eyes do tend to wander a bit.' As far as is known, these were the only Chastise crews who had the opportunity to practise with full-size dummy Upkeeps ahead of the attack, though several other pilots flew as observers in the bomb-dropping aircraft.

On 13 May, a Lancaster piloted by S/Ldr. 'Shorty' Longbottom, who had participated in many of the earlier trials, dropped a live Upkeep in the sea off Broadstairs from a height of seventy-five feet. Spinning at 500 rpm, it bounced seven times and travelled almost eight hundred yards before sinking – and exploding. Gibson flew alongside in a second aircraft, as an observer, to study the behaviour of the bomb – and to watch water spout more than five hundred feet into the air following detonation. From the squadron commander's viewpoint, this was an important exercise. He witnessed a wholly successful test of Upkeep, which had been a flight of fantasy only a few weeks earlier. He could now tell crews that they were to join battle armed with a revolutionary weapon which he knew from personal experience to be capable of achieving something extraordinary, if they carried out their allotted roles. Survivors of Chastise attested later that a critical element of Gibson's leadership was that he imbued them with his own faith. Most of his men eventually took off for Germany convinced that they could breach the dams – because he had convinced them that he himself was convinced.

By 14 May it had become evident that Maudslay's damaged aircraft would not return from the workshops for days. He was instead allotted a replacement – Z-Zebra – which the crew air-tested, and subsequently flew to the dams. HQ 5 Group reported to Bomber Command that during 617's training there had been two 'avoidable accidents'; the pilots involved were not identified, but were presumably Maudslay and Munro. The consequence for the latter, to his considerable chagrin, was demotion from the Möhne First Wave to the Sorpe Second. The obvious conclusion to be derived from what proved the last test drops of full-sized Upkeeps was that the margin of error, both for successful attacks and survival, was virtually zero.

The records do not reveal why some significant crew changes were made during the last weeks, days and even hours before the raid. Several men were removed because they were deemed incapable of fulfilling vital roles – for instance, John Hopgood's bomb-aimer. It is possible that a handful of other men decided that this terrifying, apparently almost suicidal mission was not for them. Hopgood's navigator was among those who reported sick at the beginning of May, to be replaced by a Canadian, such as was also his new bomb-aimer John Fraser – both described by the pilot as 'grand chaps'. On 2 May, Maltby's front-gunner was sacked for an unspecified disciplinary offence, serious enough to cause him to be dispatched for a spell at Brighton's so-called Aircrew Refresher Centre – a detention barracks.

Three of Maltby's final Chastise crew had never flown a single operation, while two others had experienced only a handful, which caused them to be teased by the pilot as 'sprogs' or 'rookies'. It was far from ideal for any man to embark upon the most challenging mission of his life in the company of crew members with whom he had lacked time to forge intimacy or trust. Ken Brown's front-gunner reported sick on the morning of the raid, to be replaced by a man whose pilot was unfit to fly.

A crew's absolute dependence on their helmsman was a special aspect of bomber operations. If a soldier on the battlefield recoiled from following his leader, he could and sometimes did hug the ground. But all of a Lancaster's occupants were at the mercy of the man at the controls. If he chose to be brave, they must be brave with him. If he flinched, they did likewise. If he was a poor pilot, as some were, his shortcomings could render their lives forfeit. Here was part of the unique bond – a cynic might brand it as bondage – that linked the members of a bomber crew.

In those last days, several men who had not already done so made wills, including the pilot Bill Astell, whose signature was witnessed by Henry Maudslay and Australian Norman Barlow. He bequeathed everything to his father, managing director of a Manchester textile company. 617's medical officer, Malcolm Arthurton, flew in Maudslay's aircraft on the night of 14 May for what proved their final exercise, over Uppingham and Colchester reservoirs. Among the pilots who would execute Chastise, only Martin, Maltby and Townsend were missing, for assorted technical and administrative reasons. The others were airborne for four hours, practising their approaches without dropping Upkeeps, after which Arthurton wrote in his diary: 'I have not the foggiest notion where we were nor exactly what we were doing except that we were doing low flying … People said very little and I did not embarrass them with very difficult questions as I realised there was something in the wind.' That night, Scampton's base commander Charles Whitworth flew with Gibson, who recorded in his logbook: 'Full dress rehearsal … Completely successful.' He told the crews, 'Bloody good show, boys.'

The final operational plan for Chastise reflected the primacy in British minds of the Möhne, to which Gibson himself, together with his best eight crews, were directed. This was an entirely sound call, but thereafter the targeting decisions were ill-judged. Once the Möhne was breached, Gibson's 'First Team' was ordered to move on to the Eder. There is little doubt that this was because it seemed most likely to succumb to Upkeep, despite its irrelevance to the Ruhr water system. Meanwhile the 'Second Team' was to head for the Sorpe; the

'Mobile Reserve' was expected to be ordered to the Ennepe, Lister and other masonry dams.

Just five days before Chastise was launched, Wallis submitted to Gp. Capt. Syd Bufton at the Air Ministry's Directorate of Bomber Operations a new memorandum about the Sorpe, which was clearly and rightly troubling him. Despite the tunnel vision that every visionary requires, Wallis could not ignore the overwhelming evidence that, to strike the devastating blow at German water resources of which he had dreamed since 1940, the Sorpe as well as the Möhne must be broken. He now said that, after studying aerial photographs of the dam, he recognised that its sloping face on the air side – below the reservoir – 'appears to be made of pretty heavy material', which might well hold even if the concrete core of the wall cracked.

The engineer therefore believed that it would be advantageous to attempt to create craters on the air side before dropping Upkeeps on the water side. His calculations suggested that an Upkeep dropped forty-one feet from the core would roll 113 feet down the slope into the water before exploding. If these conditions were met, he believed that a chance of 'real success' against the Sorpe still existed. Yet neither Bufton nor 5 Group made any attempt to modify the operational plan in the light of Wallis's remarks, surely because they knew that his proposal was wholly impractical operationally.

The most important reality, at this stage, was that the Royal Air Force and its senior officers – headed by Sir Charles Portal and with the notable exception of Sir Arthur Harris – were now so deeply committed to Chastise that it was unthinkable to draw back from it. Aircraft of 617 Squadron had dropped Upkeeps successfully at Reculver, demonstrating that an attack on the dams was feasible. Technical resources had been poured

into transforming Wallis's conception into reality. The bomb
dump at Scampton held fifty-six Torpex-filled Upkeeps. Guy
Gibson reported his squadron thoroughly trained for their
mission. The Air Ministry could not now abort the operation
merely because doubts persisted about whether it would
achieve its intended drastic impact on the Nazi war machine.
That view was implicitly shared by Wallis: he was too rational
a man honestly to believe that 617 Squadron had better than,
say, a 10 per cent chance of breaching the Sorpe.

What was assured, in the minds of Portal, Bottomley, Bufton
and others, was that destruction even of the Möhne alone
would constitute a dramatic, spectacular success in the eyes of
the world. It would be a fabulous act of military theatre, such
as the prime minister rightly judged indispensable to sustain
the morale of the British and indeed American peoples through
the long interlude before major land operations could be
renewed against the Germans on the continent. Even a limited
fulfilment of Chastise must chastise the enemy in the most
conspicuous fashion, enhancing the prestige of Britain and of
the Royal Air Force. And should it fail, nobody other than a
few hundred airmen need ever know about the commitment
of brains, courage, ingenuity, faith, scientific and industrial
resources that it represented. All this was valid then, and
remains so today.

One further important hurdle had to be cleared before Chastise
was launched. Through the past weeks in which the RAF had
nursed its bouncing Upkeep towards operational readiness,
parallel testing had continued on Wallis's smaller bouncing
Highball, to which the Royal Navy had cherished a more
consistent commitment than the airmen towards their own
weapon. Important admirals were anxious that no Upkeep

assault on Germany should take place before their own weapon was ready, lest it expose the secret of the technology in a fashion that must wreck the navy's hopes – the Italian fleet had been identified as a priority target. On 9/10 May, Highball trials at Loch Striven in Scotland proved an ignominious failure. Mosquito pilots performed their parts well, delivering dummy spheres that bounced across the water towards the hull of the old French battleship *Courbet*. On impact, however, these prototypes sank or were rendered inoperative. It became obvious that, whatever the long-term value of Highball, there was no early prospect of mounting Operation Servant against the *Tirpitz*. Thus the Royal Navy demanded a postponement of Chastise.

Britain's chiefs of staff were in those days closeted with their American counterparts at Williamsburg, Virginia, but before departing for the US the First Sea Lord had given instructions that his service opposed 'premature' use of Upkeep. At a meeting of the vice-chiefs in London it was agreed that only their bosses could arbitrate this vexed issue. At 6.55 p.m. on 13 May, a signal was dispatched to Washington. Next morning Portal's deputy, Air Chief Marshal Sir Douglas Evill, sent his own message to the CAS, emphasising the Air Ministry's conviction that, given the height of the German reservoirs and good weather prospects for the coming days and nights, it was now or never for Chastise. Even before this reached Portal, however, a brief signal came back from Washington at 3.55 p.m. on the 14th: 'For reasons stated by you, Chiefs of Staff agree to immediate use of Upkeep without waiting for Highball.'

At 9 a.m. on the 15th the Air Ministry sent a 'most immediate, most secret' order to High Wycombe: 'Op. CHASTISE. Immediate attack of targets "X", "Y", "Z"' – the Möhne, the Eder and the Sorpe – 'approved. Execute at first suitable opportunity.' Last-minute photo-reconnaissance by a Spitfire XI of 542 Squadron confirmed ideal water levels in the dams, together with the absence of any new defences. In the Sauerland, the 1943 spring-crop sowing weather had proved wonderfully benign: local sages were nodding wisely to each other, predicting that autumn would bring a bumper harvest. From High Wycombe, Bomber Command dispatched the order to 5 Group: 'Go.'

6

Chastise

1 TAKE-OFF

On the morning of Saturday, 15 May, Ralph Cochrane made the short journey from Grantham to Scampton to convey the news personally to Charles Whitworth and Guy Gibson that Chastise would take place on the following night, barring a last-minute emergency or weather issue. The CO of 617 was authorised to inform his flight commanders that evening. Mutt Summers arrived in a white Coastal Command Wellington, bringing from Weybridge Barnes Wallis. In the course of the day a start was made on bombing-up aircraft with Upkeeps, a laborious process for which the loading crane and a special 'dolly' load-carrier could be used on only one machine at a time. A few pilots – Barlow, Byers, Astell – air-tested their Lancasters or conducted bombing practice at Wainfleet. That afternoon Gibson accompanied Cochrane back to St Vincent's, 5 Group headquarters, for a final discussion with his staff about the Chastise operation order, No. B976.

At 1800 in Whitworth's quarters the squadron commander and the engineer met Young, Maudslay, bombing leader Bob Hay and Hopgood, whom Gibson designated as his deputy for the Möhne attack. That meeting was never chronicled, but its mood cannot have been other than tense and earnest. Hopgood successfully proposed a slight northerly change of outbound route to the Möhne, to avoid a heavily defended rubber factory at Hüls, of which he had uncomfortable experience. It is a melancholy reflection that not one of the five fliers assembled that evening would survive the war, while three had less than thirty-six hours to live.

When their meeting broke up, Whitworth took Gibson aside: 'Look here, Guy, I'm awfully sorry. Nigger's had it; he has just been run over by a car outside the camp. He was killed instantaneously.' The squadron commander's beloved dog had dashed onto the main road outside the guardroom, and paid the price. His body now lay in a detention cell. Gibson was devastated by the loss: Nigger had been his closest comrade through good times and bad, frankly a more sympathetic one than Eve. Against regulations, he had accompanied 617's CO into the mess, slept in his quarters, sometimes joined him in the cockpit. Now the dog had abruptly vanished as this lonely young man stood on the cusp of the most important deed of his life. Gibson wrote: 'And so I went back to my room on the eve of this adventure with my dog gone. I was alone looking at the scratch marks on the door Nigger used to make when he wanted to go out, and feeling very depressed.' The airman nonetheless showed his grief less than did Barnes Wallis. Although the engineer cherished a lifelong indifference to animals, through the ensuing day and night he was haunted by fears that Nigger's death was an ill omen; that it might damp Gibson's spirit at the very moment when it was supremely important that he should give his all.

Ground staff were warned that there would be a 'night flying programme' on the 16th, a fiction which was maintained in place of an 'order of battle' on squadron paperwork. This version appears to have been accepted by 57 Squadron, 617's Scampton neighbour, who had grown accustomed to mocking Gibson's men for their apparently 'cushy' war. Those who were alerted to arm and fuel the Lancasters were undeluded, however. Ground staff and aircrew alike had no notion whither their aircraft would be going come the next darkness, but they understood that they were destined at last to confront the enemy.

Adjutant Harry Humphries, who was excluded from the secret of the targets, shared quarters with P/O George Gregory, an unusual front-gunner because he was commissioned. A printer in civilian life, Gregory was a big, tough, direct twenty-five-year-old Scot who had completed a tour of operations with 44 Squadron. His wife Margaret had come down from Scotland a few days earlier, to meet him in Lincoln; it had been obvious to her that her husband was overwound – frankly jumpy. That Saturday night, as Gregory left the mess anteroom with Humphries, his pilot Hoppy Hopgood aimed a playful kick towards his backside, which caused Gregory to turn and hurl himself upon his skipper. Here was the sort of romp that 617's recent ex-schoolboys probably relished more than did older men such as Gregory. Hopgood regained his feet some-what shaken. 'Just as I said,' he told the room unconvincingly. 'Air gunners are all bloody brawn and no brains.' Back in their quarters, after a cup of tea Gregory said to Humphries, 'I think I will go to bed. May be working tomorrow.'

Sunday morning dawned bright and clear. Gibson rose at 0530, after just five hours' sleep, troubled by thoughts of his dog. He saw the medical officer to discuss his intermittently agonising foot pain, telling him that he must fly that day, but warning him to give no hint of this to others. Pilots visited the hangars and dispersals to check their aircraft, as ground crews stripped off engine covers and polished Perspex – not for show, but because in the sky over Germany smears on cockpit or turret panels could be dangerously deceptive. Compasses had to be reswung and tested with the bombs aboard, because their huge metal bulk affected needle stability. Maudslay, 'B' Flight's commander, checked Z-Zebra, then the only spare among the squadron's twenty Type 464 Lancasters. Activity of any kind was welcome to all those who would fly, as an alternative to passive reflection about what they must face, come the night. There was an alarming incident while Mick Martin and his crew were checking the cockpit of P-Popsie: somebody appears to have pressed a sensitive toggle, causing the retaining arms holding the aircraft's Upkeep to spring aside, and the bomb to thud onto the tarmac beneath. Every man fled the aircraft and vicinity, until the armaments officer arrived to assure them, with professional disdain, that the unfused weapon posed no risk.

Two crews, those of Sgt. Bill Divall and F/Lt. Harold Wilson, were stood down. Divall had suffered a knee injury, but it is unknown whether Wilson and his crew were genuinely ill, or instead considered by Gibson unfit to participate in Chastise. The name of F/Sgt. Ken Brown was belatedly added to the list of pilots who would fly. Eighteen thousand rounds of .303 ammunition were loaded into each operational Lancaster, in belts conveyed on runways to the turrets, three thousand rounds per gun. The RAF's night bombers might have fared better – would certainly have flown higher and up to 50 mph faster – without such a weight of armament. However, the hydraulic power-operated turret was among the proudest creations of Britain's pre-war aviation industry, a key element in fulfilling the air evangelists' concept of the 'self-defending bomber formation'.

Punitive losses had obliged commanders to recognise that this was a myth, forcing abandonment of almost all daylight bombing. The RAF nonetheless retained turrets in its night bombers: a ton and a half of the operational weight of a Lancaster was composed of defensive armament. Rear-gunners served a critical function as lookouts for enemy fighters, but their .303 Browning machine-guns very seldom hit anything. The most successful British bomber of the war proved to be the unarmed Mosquito, which relied on speed and height to escape enemy fighters.

The 'heavies', however, persisted with guns, as an article of faith. Commanders argued that crews valued the illusory protection, and would have felt naked without it. Bomber Command's Operational Research Section repeatedly urged stripping down the aircraft, but got nowhere. Freeman Dyson wrote: 'To push the idea of ripping out gun turrets, against the official mythology of the gallant gunner defending his crew mates ... was not the kind of suggestion the commander-in-chief liked to hear.' Chastise was one of the few operations for which the Lancaster's twin front guns would fulfil a credible function – shooting back at the flak positions on the Möhne – at the cost of pinpointing the aircraft, founts of their own streams of tracer.

The armourers required half an hour to load each four-and-a-half-ton Upkeep into the bomb bay, checking that the cylinder was securely gripped between the protruding calipers and spinning freely on its driving mechanism. Each weapon had been test-balanced, some less carefully than others: in the air, at least one crew would be alarmed by the extravagant judder of their load, some instruments becoming almost unreadable. Barnes Wallis bustled among the aircraft, checking the processes of preparation.

The first important event of the morning was a meeting of pilots and navigators, addressed by Gibson. At last, they were told the Big Thing. The mystery that had baffled them for so many weeks was resolved: their targets in Germany, which nobody claimed to have guessed already. For weeks they had feared being ordered to attack heavily defended battleships, or the Brittany U-Boat pens. 'That sounded a pretty terrifying operation,' said one man later. 'We were rather relieved when it turned out to be the dams – which nobody knew much about.'

They crowded round the models of the Möhne and the Sorpe, which for weeks had been closely guarded at 5 Group headquarters. The Eder model was not completed until the following day, a lapse that was almost certainly a reflection of its lesser significance in the minds of the planners. This proved a serious deficiency: only later that night would crews confront, and struggle to overcome, the towering topographical problems posed by the hills surrounding the Eder lake, which no aerial photo, nor even the pre-war tourist postcards that were also on display, showed in proper perspective.

The 5 Group operation order explained the industrial importance of the dams to Nazi Germany, though in surprisingly tentative phrases by comparison with the extravagant certitudes of Portal and Wallis: it suggested that destruction of the Möhne alone – 'TARGET X' – 'might well cause havoc in the Ruhr valley', adding that 'additional destruction of one or more of the five major dams in the Ruhr Area would greatly increase the effect and hasten the resulting damage ... TARGET Z (the Sorpe) is next in importance [after the Möhne].' Breaking the Eder – 'TARGET Y' – 'would seriously hamper transport in the Mitteland Canal and in the Weser (river) and would probably lead to an almost complete cessation of the great volume of traffic now using these waterways'.

Crews were warned – accurately – of three probable single-barrelled 20mm light flak guns on the wall of the Möhne, together with a further three positions below and north of the dam. There was also a possible searchlight installation, and a double-line steel-chain boom to block torpedoes, positioned some yards above the dam and supported by floating timber spreaders. An additional three dams – the Lister, Ennepe and Diemel – were identified as 'last resorts' for crews unable for any reason to reach the priority targets. These six objectives represented all but one of the seven listed by Wallis in his 1942 'Air Attack on Dams' paper.

For the attack, the squadron would divide into three waves. The First, comprising three elements of three aircraft led by Guy Gibson himself in G-George, would cross the Scheldt Estuary to attack first the Möhne, then the Eder, after which its remaining Upkeeps would be sent to address the Sorpe. It is notable that Sid Hobday, one of the navigators, always later referred to this as 'Main Force'. The Second Wave, of five aircraft, would enter Germany by a different course that crossed the Dutch coast a hundred miles further north, then take a slightly longer route to the Sorpe. Joe McCarthy and others were dismayed to be told that they would be dropping their bombs in an entirely different fashion from that which they had practised, without bounce. The American acknowledged later: 'We weren't too happy about it ... We studied the models of the dams and I could see I was going to have a problem with mine.' The remaining five aircraft would take off much later, and form a Mobile Reserve, under the direct wirelessed orders of 5 Group, to attack whichever dams remained unbreached by the time they were airborne over the continent. All Lancasters circling the targets were instructed to fly anti-clockwise patterns.

Crews' fears, at this stage, focused heavily upon the routes they would fly to the targets. Gibson had earlier visited Tempsford in Bedfordshire to discuss this problem with Charles Pickard, the officer commanding the Special Duties squadron based there, whose aircraft flew constantly at low level over Europe to rendezvous with Resistance groups. Pickard's views influenced 617's choices of coastal crossing-points, but in May 1943 there was no safe, secret path into Germany. At extreme low level, the dam raiders must pass close to some of its heaviest defences. Visually-sighted light automatic weapons could inflict carnage at close range upon unarmoured Lancasters. The staff had selected multiple dogleg flight plans that should enable the squadron to weave around the most dangerous flak belts. It was nonetheless obvious that, whatever later took place at the dams, merely getting to them posed exceptional challenges.

When the aircrew dispersed after three hours' debate, they were warned to avoid any discussion of the operation around the airfield ahead of the full squadron briefing that evening: there must be no gossip with ground personnel. Gibson wrote: 'The rest of the day was a terrific flap.' He himself became involved in a petty squabble which re-emphasised his unpopularity with those whom he deemed small fry, including erks. He ordered a member of the station workshop staff, F/Sgt. Brown, to make a coffin for Nigger. The man refused: such a task had nothing to do with his duties. Gibson lost his temper. High or low words were exchanged, which ended with the squadron commander departing, without a coffin, to instruct Sandy Powell to bury the dog outside his office at midnight that night. Powell assented, but in the event delegated an airman to carry out the task.

Shortly before noon, wireless-operators were briefed by 5 Group's signals officer about procedures and codewords. Silence would be preserved on the ether until they were over the targets, but a listening watch was especially important for the Reserve Wave, who would be directed towards objectives only while crossing Holland. Gunners and other specialists likewise received their own briefings, and studied the target models and photos. Twenty-year-old Canadian Fred Sutherland, a doctor's son from Peace River in Alberta, gazed in dismay at the Möhne in the smoke-filled briefing room: 'I immediately thought we didn't have a hope,' he said later. It was plain that 'this was going to be really touch-and-go'. Meanwhile Barnes Wallis became distressed by a last-minute technical glitch: the wrong oil was apparently being used on some Upkeep equipment; the right stuff was eventually located.

Before the final briefing, navigators pored over their maps, marking flight plans and courses, laying thick red crayon marks on power lines, mortal hazards for low-fliers. Most aircrew returned for a while to their quarters, some to fulfil the familiar ritual of putting out Last Letters for loved ones, to be once again locked away on their return – or not, as events might decide. Most such missives were composed not to lay bare the thoughts and fears that filled men's hearts, but instead to comfort the bereaved, like that composed by Maltby's wireless-operator Sgt. Antony Stone, who wrote for his parents in Winchester that if he perished over Germany, 'I will have ended happily, so have no fears of how I ended as I have the finest crowd of fellows in the world with me, and if the skipper goes I will be glad to go with him. He has so much to lose, far more responsibilities than I.' Here, Stone was obviously thinking of Maltby's pregnant wife. Charlie Williams, Norman Barlow's Australian wireless-operator, wrote to his adored fiancée Bobbie, a secretary in Nottingham, discussing wedding plans. 'Cheerio for now darling,' he signed off, 'and believe me when I say that I love you very dearly and always will.'

Some wrote in the exalted manner of an airman on another squadron, a twenty-five-year-old former preparatory school teacher who told his parents that the war 'has shown me new realms where man is free from earthly restrictions and conventions, where he can be himself playing hide-and-seek with the clouds, or watching a strangely silent world beneath ... So please don't pity me for the price I have had to pay for this experience ... Now I am off to the source of Music.'

Yet another RAF flier's last words to his widowed mother became famous in wartime Britain: 'I still maintain that this war is a very good thing; every individual is having the chance to give and dare all for his principles like the martyrs of old. However long the time may be, one thing can never be altered – I shall have lived and died an Englishman. Nothing else matters one jot ... I count myself fortunate in that I have seen the whole country and known men of every calling. But with the final test of war I consider my character fully developed. Thus at my early age my earthly mission is already fulfilled and I am prepared to die ... you will live in peace and freedom and I shall have directly contributed to that.' To a later generation such sentiments may sound mawkish, yet they were sincerely held by many young men of that time and place; were indeed indispensable, to arm them with the courage to set forth upon such an 'op' as Chastise.

At 1800, all 133 men who were to fly that night assembled in the big briefing room on the upper floor of Scampton's sergeants' mess, seated crew by crew on lines of benches. The hum of chatter was stilled, and the crews rose to their feet as

Cochrane, Whitworth, Gibson and Wallis entered with a tail of staff officers and a man from the group's meteorological office. Wallis's assistant Herbert Jeffree also slipped in. On the dais at the far end of the room stood a blackboard, beside it a large map of Europe on which their routes were marked with red tapes. Gibson gazed at his men with a pride that was wholly justifiable, reflecting upon what he had made them in a few short weeks: 'rather tousled and a little scruffy, and perhaps a little old-looking in spite of their youth. But now they were experts, beautifully trained, and each one knew his job as well as any man had ever known any job he was to do.' This was an overstatement: some of his men would struggle, and indeed die, that night, because Chastise demanded from them more than they were capable of giving. But it is a deeply moving aspect of the operation, that in an age when the concept of duty meant much to many people, the aircrew of 617 would strive to the limits of their powers, and beyond, to do what was now to be asked of them.

The CO began by introducing his principal guest, the white-haired, studious-looking engineer. Wallis took the floor, and staged one of the performances of his life. This naturally gentle man assumed the fervour and conviction of a missionary priest as he explained the importance of Germany's dams: the Möhne reservoir contained approximately 130 million tons of water, the Eder 202, he told them. The former must be something special, he said, because thirty years earlier the Kaiser had opened it personally. With the aid of blackboard and chalk, he explained the evolution and workings of his extraordinary weapon. The pilots and navigators already knew some of this from their own earlier briefing, but nobody afterwards confessed to boredom. They were hearing of something so remarkable, so far beyond their experience and – for almost all those present – a literal matter of life and death, that Wallis was never in danger of losing his audience. Navigator Sid Hobday said: 'We were very impressed with him and thought he was a marvellous man.' Bomb-aimer Jim Clay thought it incongruous that such a gentle, kindly-looking figure 'should be involved with devastation'.

When Wallis sat down Cochrane rose, to assert his absolute confidence in the mission's success. He added that he believed it would become historic, but warned with his accustomed chill: 'Don't think you are going to get your pictures in the papers.' Upkeep, he said, would be used on other operations in the future. Its secrets must be kept until the war ended. This reflected an uncharacteristic naïveté on the part of the AOC, for it made the large assumption that the Germans would be denied a close look at the aircraft which executed Chastise, and indeed at an Upkeep, before the night was out.

Syd Bufton had already warned High Wycombe that any or all information about the raid following its execution would be released to the BBC and the newspapers by the Air Ministry in King Charles Street, and not by Bomber Command. The RAF's most senior officers cherished hopes of exploiting a huge propaganda opportunity if Chastise succeeded. If it failed, they had both public and private motives for seeking to conceal from the enemy, and from the world, how much technological effort had been committed to it, albeit only a handful of aircraft.

In the briefing room at Scampton, after the weatherman had given the night's forecast and – most significant for navigators – the anticipated wind speeds, it was again Gibson's turn, though he had already been on his feet – those painful feet – for hours that day. He rehearsed once more the operational details for the successive waves, warning that light flak would pose a threat throughout their flights to and from the target.

Accurate map-reading was vital; sticking tightly to routes and turning points that had been painstakingly plotted to sidestep known enemy concentrations, though there was always the risk of newly deployed, and thus unidentified, batteries. Then the meeting broke up, leaving every man thoughtful, some more bullish than others.

At 1930 the fliers adjourned either to officers' or sergeants' messes for the usual pre-flight meal. Security had billed that evening's activity as merely another round of training, but those with eyes to see noticed that bacon and eggs were being served, luxuries in wartime Britain such as were readily accessible only to men not unlikely to be dead before morning in their country's service. Wallis had a last exchange with the squadron commander, in which he expressed the fervent hope that 'Gibby' and all his men would come home. The engineer recalled Harmondsworth and Nant-y-Gro, an age ago, then added: 'I look upon this raid as my last great experiment to see if it can be done on the actual thing.'

Cochrane was considerate enough to visit the hangars and address some of the erks who had laboured almost around the clock to prepare aircraft and bombs, and to overcome last-minute problems. He said: 'I realise how difficult it has been for you and the work you have had to put in for all of this, and I just hope that in the morning you'll consider it has all been worthwhile.' His audience of fitters and riggers still had no idea what 'their' Lancasters were going to do. When the station medical officer tried to telephone a member of his family who was sick, he discovered from the switchboard that a ban on outside calls was being enforced.

At 2000, ninety minutes before the first take-offs, Gibson's men drifted towards the crew rooms beside the squadron's hangars, some on foot and some on bikes, to strip themselves of personal possessions, don flying kit, collect parachutes, flight bags and flying rations – chocolate, sandwiches, fruit juice, perhaps an orange and chewing gum. They were issued escape equipment, including Dutch and German currency, miniature compass, fishing line, silk maps.

Then they lay on the grass chatting, smoking, nursing private thoughts in the beauty of the Lincolnshire evening. Gibson described his own sensations at such moments before setting forth on an operation, when almost every flier's anticipation was at its most acute: 'Your stomach feels as though it wants to hit your backbone. You can't stand still. You laugh at small jokes, loudly, stupidly. You smoke far too many cigarettes, usually only half-way through, then throw them away. Sometimes you feel sick and want to go to the lavatory. The smallest incidents annoy you and you flare up on the slightest provocation ... All this because you're frightened, scared stiff.' Even in Gibson, fear was there. He had learned how to conquer it, however, and sometimes showed himself harshly unforgiving towards others who were less fortunate.

For all the breezy confidence that most men exuded when among a crowd, F/Sgt. Bill Townsend, a veteran of twenty-six operations, confessed to feeling sick, convinced that they were all 'for the chop'. John Hopgood told Dave Shannon as they smoked behind the hangar, 'I think this is going to be a tough one, and I don't think I'm coming back, Dave.' This shook Shannon, who urged his friend that having 'beaten these bastards' for so long, he could do it again. Lewis Burpee, the musician's son who knew that his wife was expecting their first baby, shook the hand of fellow-Canadian Ken Brown, saying 'Goodbye Ken' with undisguised finality. Brown's rear-gunner spoke confidently about the occupants of other aircraft who shared their bus to the dispersals: 'You know those two crews aren't coming back, don't you?' Many men of Bomber Command, on many nights, experienced such premonitions, and most went unfulfilled ... until the last time.

At 2030, Gibson gave the word: time to go. Crew by crew, the men of the early waves clambered aboard the buses and trucks that bore them the few hundred yards to the Lancasters, where each pilot was handed by the ground-crew chief a clipboard bearing Form 700, to be signed in token of acceptance that his aircraft was fit and ready to operate. Sgt. Abram Garshowitz, one of twelve children born to Russian immigrants in Hamilton, Ontario, chalked on the Upkeep of Bill Astell's B-Baker, of which he was wireless-operator: 'Never has so much been expected of so few.'

Then each crew clambered up the short ladder to the rear fuselage door, swung themselves over the bulk of the main spar into the cockpit, and settled in their respective positions. Navigators laid out maps; gunners took the cramped seats behind their weapons, trained by twist-grips to which triggers were fitted. Each aircraft comprised an all-up weight of sixty-three thousand pounds, of which the crew accounted for 1,400 lb, turrets and ammunition over two thousand, Upkeep 9,250 lb, together with 1,740 gallons of fuel and 150 gallons of oil.

At 2100 a red light soared from the Very pistol of Gibson's G-George, fired by his wireless-operator Bob Hutchison, to signal the First and Second Waves to start engines. Each pilot ran through the checklist, repeated back by the flight-engineer.

'Switches off.'

'Inner tanks on.'

'Immersed pumps on.'

'Check seat secure.'

'Brakes on and pressure up.'

'Undercarriage locked.'

'Flaps Thirty.'

'Radiators closed.'

'Lock throttles.'

'Prepare to start up.'

Erks had primed the engines. The pilot called down through his window, 'Contact-starboard inner!' then continued through the ritual until each of the four Merlins was roaring throatily, like a battery of gigantic lawnmowers. Joe McCarthy was exasperated to discover a coolant leak on a starboard engine. His aircraft, Q-Queenie, was thus unfit to take off, just as he and his crew prepared to participate in the most important event of their lives. They clambered down, and dashed for the newly-arrived T-Tommy, which had been flown up to Scampton from Boscombe Down a few hours earlier, after being used for Upkeep experiments; it had since been armed and fuelled for just such an emergency as this. The big, fair American began rattling through the cockpit checks for the second time that evening – then exploded anew on finding that there was no compass-deviation card, essential to set a course. He jumped to the ground, ran back to the truck that had brought them from the flight offices, and hastened to accost the adjutant Humphries, cursing freely: 'Where are those lazy, idle, incompetent compass-adjusters?'

A clutch of ground staff dispersed in search of a card while Humphries sought to calm the sweating, furious pilot, who kept clenching and unclenching his huge hands in frustration. At last Chiefy Powell ran up with the vital accessory. McCarthy grabbed it in one hand, his parachute in the other, and raced back to board T-Tommy. But he carelessly seized the ripcord rather than the handle of the parachute, which suddenly burst open, spewing silk across the grass. He hurled it away in disgust, leapt up the ladder into the aircraft without breaking stride, and started the engines within five minutes – somebody pushed another parachute aboard just before T-Tommy began to taxi.

The other eighteen aircraft started up without trouble, however. Gunners swung their turrets, checking the hydraulics; hatches were reported 'secure and locked'. However immature and unruly might be some aircrew on the ground, once embarked upon an operation every man who wished to survive became meticulous. Word was passed down through the cockpit window: 'Chocks away.' An airman on the ground gave a thumbs-up, signalling that the wooden wheel stops had been dragged clear, followed by a hiss of air as brakes were released. One by one the big aircraft began to bump across the field. Scampton's grass was anything but ideal for Lancasters as heavily laden as those of 617 that night, which was why it was scheduled for concreting. The spring was dry, however; take-offs should be safe on the south-west/north-east axis usually favoured as offering the longest run.

In the usual course of the bomber war, across eastern England on an operational night hundreds of aircraft would have been getting airborne alongside those from Scampton. On 16 May, however, the moonlight meant that Bomber Command's front line was 'stood down', allowing crews to savour their night's reprieve in Nottingham's Trip to Jerusalem, Betty's Bar in York, the Black Bull in Lincoln and a hundred other such hostelries. Only a few bombers – two Lancasters, thirteen Stirlings and thirty-nine Wellingtons – were committed to sea-mining sorties off the continental coast. 'Gardening', as it was known, was an unsung success of Bomber Command's war, responsible for sinking over half a million tons of enemy shipping, and much less hazardous than attacking Germany.

Such missions were thus often assigned to novices graduating from operational training units: on the night of 16 May just one minelayer was lost, off Brest. Meanwhile three Mosquito 'nuisance-raiders' attacked Berlin to deprive its inhabitants of sleep; two did likewise to Cologne, two to Düsseldorf, two to Münster. Four Wellingtons dropped propaganda leaflets over Orléans. Otherwise, in the midst of those years in which the skies over north-west Europe were habitually torn asunder by dramas, tragedies, fire and destruction, through the hours that lay ahead 617 would have all to itself the darkness – and the enemy.

On the grass at Scampton, at 2128 the first aircraft of the Sorpe Wave, E-Easy, tested its engines to full power, checking magnetos and airscrew pitch control. The pilot spoke over the intercom: 'OK, everybody? OK behind, rear-gunner?' Then a green Aldis lamp flashed from the little caravan beside the runway. The pilot said 'Full power,' the flight-engineer repeated his words and opened the throttles. Brakes were slipped and the Lancasters lumbered, lurched, then surged across Scampton's huge field, slowly gathering speed. That evening, aircraft seemed to take an eternity to lift off: the weight of their Upkeeps caused them to hug the earth until the boundary hedge was rushing towards them at alarming speed. A navigator said: 'I had visions of the bumpy take-off causing the lights under the fuselage to be shaken off, so that instead of being sixty feet above the ground we would finish up sixty feet underneath it.' At last, however, the aircraft unstuck. Each crew in turn heard the welcome words exchanged over the intercom in the cockpit: 'Climbing power … Wheels up … Flaps up … Cruising power.'

E-Easy was followed at one-minute intervals by Les Munro in W-Willie, Vernon Byers in K-King, Geoff Rice in H-Harry.

These aircraft of the 'Second Wave' took off first, because their route was longer, and they were to fly and attack independently. Gibson, Martin and Hopgood then took off together at 2139 in G-George, P-Popsie and M-Mother, presenting a striking spectacle, because it was rare for Lancasters to fly in threes. Some of the ground staff were too tired to stay awake to watch this historic moment: ground fitter Ken Lucas was one of the team that had laboured until 0300 that morning to repair damage suffered by aircraft on the Reculver bomb runs. Now he lay slumbering on his bunk in the old station married quarters.

Gibson's trio were followed eight minutes later by Young, Maltby and Shannon, respectively in A-Apple, J-Johnny and L-Leather. At 2159 Maudslay, Astell and Knight took off in Z-Zebra, B-Baker and N-Nuts, followed at 2201 by Joe McCarthy in T-Tommy – twenty minutes behind schedule. The watching Harry Humphries thought, with some dismay, 'He may make a mess of the whole thing, taking off in such a state.' Once airborne, McCarthy drove his aircraft ruthlessly to make up lost time.

Some of the Reserve Wave crews watched almost in disbelief as the first planes lifted off 'with this enormous thing, almost like a garden roller, hanging underneath'. Then, with time to kill, several settled down to play poker. Sgt. Doug Webb, a twenty-year-old Londoner who was Bill Townsend's rear-gunner, took a bath, because he wanted to die clean. All five crews – Ottley in C-Charlie, Burpee in S-Sugar, Brown in F-Freddie, Townsend in O-Orange and Cyril Anderson in Y-York – eventually took off in full darkness, at one- or two-minute intervals, just after midnight. Harry Humphries and WAAF intelligence officer Fay Gillon returned to the officers' mess. The latter knew the target; the former, to his

chagrin, was still as ignorant as almost all the ground personnel about where the squadron had gone. Gibson and 132 men
whom they had come to know – not intimately, but as well as
people ever knew each other in the circumstances of the
bomber war – had vanished into the night. Those left behind
at Scampton had a dim sense of the parallel universe into
which the Lancasters had vanished, where crews faced flak,
searchlights, and a lottery in which graves were the destiny of
losers. But only the fliers now committed to Chastise could
comprehend the weird, terrifying and – for its appointed
victims – ghastly experience to which Barnes Wallis, Sir
Charles Portal and Ralph Cochrane had committed them.
Gillon said to the adjutant, 'Isn't this exciting, Humphy?'

2 GETTING THERE

Gibson's section crossed the Suffolk coast at Southwold at
2229, forty minutes after take-off. Its three Lancasters then
flew onwards in a fashion few had ever experienced, so low
that the rushing aircraft shapes were reflected on the sea
beneath them. The accustomed pattern of night bomber
operations, the received wisdom of their war, was that height
conferred improved prospects of survival, so that all the way
to Germany pilots clawed at the sky, welcoming every few
extra feet of altitude. Yet now this squadron, for this one
mission, was flying the entire route to the target below the five-
hundred-foot threshold at which German radar must pick
them up. Cochrane's planners believed, undoubtedly correctly,
that on a moonlit night which would be suicidal for Main
Force bomber operations, only a low-level approach might
spare Gibson and his squadron from slaughter by night-
fighters. The acknowledged rule of the game was that once a

fighter engaged a victim, unless an alert gunner cried a warning, precipitating a drastic 'corkscrew' descent, the bomber and its crew were probably dead meat. Only a minority of aircraft survived attack by a Ju88 or Bf110, and there could be no corkscrewing from an altitude of less than a hundred feet.

Yet while most of Goering's night-fighters stayed on the ground until scrambled following a radar alert, ten thousand flak guns of all calibres were continuously manned. Concentrations of 88mm heavies, Flak 30 and Flak 38 automatic weapons were clustered in belts between the Dutch coast and the dams. The lighter guns had no need of radar to alert them – they could hear heavy bombers coming. And whereas USAAF B-17 Flying Fortresses were heavily armoured for daylight operations, British bombers relied on darkness for protection. The Lancaster was a superb aircraft, but the price of carrying its impressive bombload was that its airframe could withstand much less punishment than its American counterparts from cannon or machine-gun fire.

That May night was full of dilemmas for the attackers. The lower they flew, the better their chances. Yet below a hundred feet their Gee electronic navigation aid, normally reliable and unjammed on the North Sea crossing, became erratic because of the curvature of the earth. Accurate course monitoring was vital to avoid the flak. On 16 May, this depended upon the skills of individual navigators, and their diligence in monitoring winds over the North Sea by dropping flame floats. Some Chastise crews suffered severely for lack of Gee assistance, succumbing to unnoticed drift caused by unforecast wind.

Gibson's three lead aircraft approached Holland at 2250. In G-George, Richard Trevor-Roper, who had stood behind the cockpit for part of the run across the North Sea, removed his

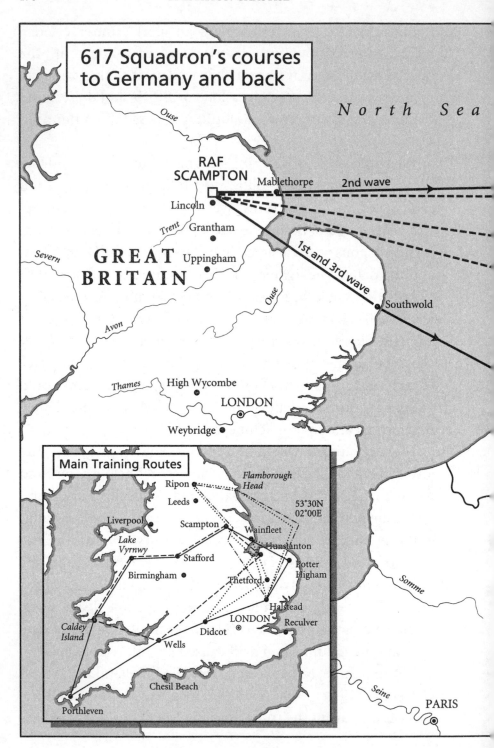

Mae West and disappeared to the rear turret. While they were still over the water John Pulford momentarily flicked on the spotlights under the plane's belly, to check their height. The beams seemed a distance from convergence, and the flight-engineer suggested dropping down a little. In the nose 'Spam' Spafford said, 'There's the coast,' to be sharply contradicted by Gibson, who said – wrongly – that it was merely low cloud and shadows on the sea. Navigator Terry Taerum left his curtained table and stood behind Gibson and Pulford, peering out of the huge greenhouse of a cockpit in hopes of a landmark. 'Can't see much,' he said, but he thought that they were on track.

He was not quite right: they had been pushed south by the wind. Micky Martin had periodically climbed to take Gee fixes, without noticing the drift. Instead of entering the mouth of the East Scheldt, they were hurtling towards the heavily defended island of Over Flakee, and it was too late to veer away. 'Stand by front-gunner,' said Gibson. 'We're going over … No talking. Here we go.' It was surely a tactical error to fire at the ground defences. The Lancasters' Brownings were most unlikely to damage the enemy, but wasted ammunition they might need later. Several German witnesses, including those at the Möhne, noted that the bombers' tracer assisted them to track the attackers.

The ground defences, surprised, stayed mute as the First Wave of bombers swept overhead. Gibson briefly climbed to three hundred feet, to give Gee a chance; Taerum and Spafford seized the opportunity to spot a landmark. Within seconds they identified a windmill and communications mast near Roosendaal, and fractionally adjusted course as the Lancaster dropped back towards 'deck level'. In the nose, Spafford began to check their progress on his map. Fleeting glimpses of houses, factories, railway lines were punctuated at intervals by an intercom yell of 'Pull up!' from the nose, signalling another power line. The lead element of the First Wave had passed through the early hazards.

Others, however, fared much less well. At almost exactly the same moment that G-George crossed the coast, 130 miles further north the Second Wave also approached land. Unlike Gibson's crews, which were in close formation, Les Munro and his fellow pilots were each making their own ways, albeit following identical courses. With visibility not much above a mile, even in the moonlight they could not see each other. Munro had constantly checked his drift over the sea by releasing flame floats. He realised that the northerly wind was stronger than forecast, and adjusted course to compensate, as some others did not. Now he saw waves breaking on the offshore island of Vlieland, a hundred feet below.

As they breasted its sand dunes, he glimpsed a sudden eruption of flame to the south-east – almost certainly an aircraft exploding. Another Lancaster of his wave – K-King, flown by Vernon Byers – had drifted over Texel, which was heavily defended. The twenty-three-year-old Canadian was the least experienced of 617's pilots, having flown just five operations before being drafted to Scampton. About 2257 his aircraft plunged into the waters of the Waddenzee, becoming the first to be lost executing Chastise. At the same moment, upcoming tracer broke the darkness below Munro – shells from a battery of Marine Flak 246, recently deployed on Vlieland, unnoticed by British reconnaissance aircraft and photo interpreters. There was a cracking noise behind the cockpit. The crew intercom went dead, severed by a cannon shell. Deprived of communications, Munro banked over the Waddenzee and shouted to Frank Appleby, his flight-engineer, to tell wireless-operator Percy Pigeon to explore their damage. After three or four minutes' circling that seemed to the crew interminable, Pigeon clambered into the cockpit and bellowed into Munro's ear above the roar of the engines: there was a big hole in the rear fuselage; the master compass was shattered; none of the crew was injured, but communications were irreparable. Deprived of an instantaneous link between bomb-aimer, navigator and pilot, it would be impossible effectively to attack the Sorpe, even if they could find their way there. Munro made the bitter, inescapable decision to turn back for Scampton.

An Early Return in the face of mechanical failure or damage was among the most painful calls a bomber pilot could make: it might expose him to the charge – sometimes valid – of faint-heartedness. In this case, however, it was obvious that Les Munro, an able and determined flier, had no choice. To have flown on towards the Sorpe merely to make the gesture of releasing an unguided Upkeep would have been futile, probably also suicidal.

Even as Munro's W-Willie turned west for home, a third member of their wave, Geoff Rice, met disaster. A twenty-five-year-old grammar-school boy from Hinckley, Leicestershire, Rice had worked in the local hosiery trade before joining the RAF in 1941. Among the least experienced Chastise pilots, he had completed just nine operations in three months with 57 Squadron when he was unwillingly drafted to 617, along with the rest of Dinghy Young's flight. Now, his H-Harry approached Vlieland so low that its altimeter was reading zero, and he struggled to judge his height above the reflections from the sea. At 2300 the aircraft suffered a sudden violent jolt, followed by a second tremor. Salt water sprayed into the cockpit, deluging the navigator and his table. From the rear turret, shocked gunner Steve Burns exclaimed, 'Christ! It's wet at the back.

You've lost the mine!' Their bomb was gone, ripped from its caliper arms by collision with the sea at over 200 mph. The aircraft's tailwheel, having been struck by the falling steel monster, had been punched up into the fuselage, where it smashed the Elsan and soaked Burns with water and disinfectant. The crew experienced a surge of shock and terror, surely matched by rage towards the pilot whose momentary loss of concentration or judgement had almost killed them. As Rice climbed, water poured out of the fuselage. Then came a surge of relief, that they had survived. They too turned for home, marvelling that H-Harry was still airworthy after its masquerade as a flying-boat.

Just two of the five aircraft in the Second Wave were still armed and heading for the Sorpe – Norm Barlow's E-Easy, and far behind him Joe McCarthy's T-Tommy. Less than two hours after the first aircraft had taken off, as Gibson's trio completed their own North Sea crossing, and happily unknown to him, Chastise had already suffered a series of crippling blows.

For crews on Main Force operations, flights towards Germany sometimes took on the monotony of an interminable progression along a blacked-out motorway. More than a few fliers smoked through a corner of their oxygen masks; some pilots were insouciant or indeed reckless enough to set their autopilots; wireless-operators sometimes tuned to a music station; a gunner occasionally dozed, though if he was caught, crewmates harrowed him mercilessly. The men flying to the dams, however, were not tempted for a moment to relax vigilance in checking their course, watching for fighters or natural hazards, anticipating the ordeal that awaited them in Germany's Sauerland. In O-Orange, George Chalmers, like many wireless-operators, spent most of the trip peering through the Perspex astrodome, looking for trouble. They were observing wireless silence as far as the target, though he kept the headphones on his ears, to receive any message from 5 Group.

Gibson's three Lancasters spotted the Wilhelmina Canal, and followed it eastwards for a few miles until north of Helmond they identified another pinpoint – a T-junction of the canal, clearly visible below – at which they turned north-east on a bearing of 061 degrees, heading for a bend in the Rhine near Rees. Ten minutes behind Gibson, Young, Maltby and Shannon made their landfall at the East Scheldt estuary exactly as planned – they had corrected drift over the North Sea by taking successful Gee fixes. Unlike the first three Lancasters, they met fierce light flak, with tracer and search-light beams slashing open the darkness; none was hit, however. As they powered on towards Germany, they continued to climb to take fixes until German jamming became effective.

Three minutes after Young's section crossed the coast, at 2313 the lonely McCarthy did likewise, a hundred-odd miles further north, having made up a little time over the North Sea, but his radio receiver appears to have suffered technical failure. He met some flak over Vlieland: 'Very hot reception from natives when crossed the coastline. They knew the track we were coming in on' – because the others had passed minutes earlier – 'so their guns were pretty well-trained when they heard my motors. But, thank God, there were two large sand dunes … which I sank between.' It was a tribute to McCarthy's determination that he kept going towards the Sorpe after encountering a succession of technical failures. Among the highest terms of praise in Bomber Command's lexicon was to call a man 'a press-on type'. The American was a press-on type. His crew's passage to the Sorpe continued to be fraught with incident, sometimes gratuitous. South-east of Hamm, front-gunner Ron Batson spotted a goods train and begged, 'Can I have a go, skip?' The pilot assented, Batson blazed at the train, and T-Tommy received back a cannon shell in the wing.

At 2321 Maudslay, Astell and Knight reached the East Scheldt estuary dead on course and only fractionally behind time. As they followed the First and Second Waves towards the Möhne, they suffered some hair-raising close calls with power lines, but otherwise escaped harm. Further north, Norm Barlow, a thirty-two-year-old garage owner with a colourful past back home in Australia, was less fortunate. At 2350, he or his bomb-aimer suffered a split-second loss of concentration, and E-Easy flew headlong into a 100,000-volt overhead cable just north of Haldern in Holland. The aircraft plunged into the ground, a mass of flame. All seven crew perished instantly, including Queensland-born wireless-operator Charlie Williams, who had written to his fiancée Bobbie in Nottingham just before take-off: 'Cheerio for now'. McCarthy's T-Tommy thus became the only Lancaster of the Sorpe Wave that was still airborne, bound for the dam.

As Gibson's lead element passed over the German frontier, Terry Taerum, son of a Norwegian father from Milo, Alberta, and a week short of his twenty-third birthday, said in his usual calm, flat tones, 'The next thing to see is the Rhine.' They found north-west Germany briefly tranquil as they surged onwards, fields and villages flashing beneath the Lancasters. Then they saw before them the mighty Rhine, and knew that the industrial conglomeration of the Ruhr lay beyond, with countless German eyes watching and reporting their progress; air-defence headquarters puzzling over their significance and the destination. Spafford identified a landmark and said, 'We are six miles south. Better turn left, skip. Duisburg is not far away.'

They banked slightly to starboard, Hopgood and Martin following suit: they were dangerously close to one of the densest flak belts in the Ruhr. Gibson asked the navigator sharply, 'How did that happen?' 'Don't know, skip. Compass u/s [unserviceable]?' 'Couldn't be.' Taerum came back, apologetic: 'I'm afraid I misread my writing, skip. The course I gave you should have been another ten degrees to port.' 'OK, Terry. That might have been an expensive mistake.'

Then the sky erupted as searchlights and light flak raked the darkness around and above them. For brief seconds they were caught in beams which prompted Trevor-Roper to open a furious fire from the rear turret. Spafford yelled from the nose that they were almost touching the crops in the fields beneath. Even Micky Martin, prince of low-fliers, was holding P-Popsie a few feet above them. In the glare of the beams Gibson's crew could read the identification letters of the Australian's plane, 'AJP', and that of Hopgood, 'AJM'. Then they were out of the lights, back in darkness, again racing over silent countryside.

South-west of Dülmen they encountered more light flak, which appears to have hit Hopgood's port wing, and possibly done worse things, including damaging an engine and inflicting some injury on the pilot, though even when they began to use the VHF voice-radio link, M-Mother's pilot said nothing to suggest that he or his aircraft was in trouble. Gibson told Hutchison to break radio silence and report the flak concentration to 5 Group, which four minutes later retransmitted a warning to all 617 aircraft. They were flying so low that Hopgood's aircraft passed beneath one set of power lines, and all three were now depending entirely on compasses, maps and visual observation to hold their course, because jamming had blinded Gee. It is unsurprising that Gibson's section veered off track several times during the passage; more remarkable that the crews proved able to correct errors after briefly losing each other, and to reach their objective almost simultaneously. The First Wave identified their last turning point at the small town of Ahlen, just nineteen miles from the Möhne. There they swung south-south-east, on a course of 196 degrees, for the last, mercifully uneventful six minutes of the flight.

Properly to grasp how Chastise unfolded, it is necessary to see the operation not as a single saga, but instead as a series of linked ones involving different groups of people. The flight to the dams was a remarkable affair in its own right, demanding the highest skill, courage and luck, even before the Lancasters launched their attacks. Already for some crews – those of Barlow, Rice, Munro, Byers – fate had run against them. The strain for the remaining fifteen was immense, of dodging flak, becoming briefly lost, recovering landmarks, climbing in pursuit of a Gee fix, flying thirty tons of metal, Perspex, fuel, armament and human flesh as if the Lancasters were stunt planes. In the second element of the First Wave, Melvin Young stayed high, up to five hundred feet, which assisted his navigator by enabling him to use Gee, while upsetting Maltby and Shannon, who hung low: they believed that Young's tactic put them all at risk.

In the rear section of the leading wave, Bill Astell, a pupil of Henry Maudslay's at HCU, had for some time been trailing his companions. Just south of Borken, after Maudslay and Knight had yet again skipped power lines, B-Baker collided with the cables head-on in a cascade of flashes, flame and giant sparks. A year earlier the pilot had survived a crash-landing after being shot down over North Africa, then walked back to base across the desert, exploiting his colloquial German. This time he was less fortunate. His blazing plane fell in nearby farmland, where its exploding ammunition provided a firework display until drowned out by the thunderous detonation of the Upkeep. The explosion, triggered by a self-destruct pistol, shattered every window in the farmhouse where a terrified family had been sleeping, and left wreckage in which reposed the broken remains of seven airmen, including Abram Garshowitz, who had chalked ironic words on their bomb about 'the few'.

At that moment, an aircraftman named Munro was completing the digging of a grave outside the squadron commander's office at Scampton. Gibson had asked Chiefy Powell to bury Nigger there at midnight, when he himself expected to be facing death above the Möhne. Munro completed his task just as the silence of the darkness at Scampton was broken once more for a take-off: the last aircraft of the Reserve Wave, Y-York, flown by Cyril Anderson, was setting forth for Germany. Anderson had served with RAF ground crew since 1934, and after retraining as a pilot flew only five operations before being posted to 617 from 49 Squadron, for reasons never explained. He was acutely conscious of his inexperience, which imposed a responsibility he should not have been asked to bear. Moreover, a few weeks earlier he had suffered a personal tragedy when his baby son died at the age of four months.

Thus, fourteen Lancasters of 617 were now headed towards the dams, with more than two hours between first and last. Five were gone before the first blow was struck. This was no direct fault of Barnes Wallis or of Upkeep, but was instead a reflection of the huge difficulties of conveying the bombs to their targets in conditions which would allow them to be used. Cochrane wrote later in his report on Chastise: 'It is well worth considering whether this operation would not have been at least equally effective and cost less if all the aircraft had flown above the effective accurate range of light flak and machine-gun fire ... On balance, for any future similar operation, it would probably be better for aircraft to fly at 2,500–3,000 feet ... and be prepared to get down to ground level if attacked by fighters.' This seems mistaken. Once German fighters had been alerted to 617's presence, and vectored towards elements of the attacking force, as they surely would have been, the consequences for Chastise's Lancasters would have been as grievous as, or worse than, those of their low-level passage. There was never an easy route from Lincolnshire to the Möhne: Gibson's aircraft were obliged to brush the Ruhr, the most heavily defended region of Nazi Germany, in order to reach its water supplies in the rural Sauerland beyond, which the leaders now glimpsed for the first time.

7

At the Dams

1 THE MÖHNE AND THE SORPE

The first German local sighting of the Möhne attackers was made by an air-raid watcher on a so-called Bismarck tower set atop a hill four miles from the dam. This architectural oddity, one of some 240 around Germany, had been intended to be surmounted by a statue of the Iron Chancellor, but his anti-clericalism made him unpopular in the Catholic Sauerland, and when he fell from power in 1890, work stopped. On the night of 16/17 May, however, the stump provided a vantage point for two local men – counterparts of Britain's Observer Corps – who hastily telephoned air-defence headquarters.

A minute after the shadows of the first Lancasters flashed over the tower, at 0015 Mick Martin was first to glimpse the Möhne, followed seconds later by Guy Gibson, who wrote: 'It looked squat and heavy and unconquerable … grey and solid in the moonlight as if it were part of the countryside itself and just as immovable.' Forty-two minutes elapsed between Martin's arrival over the Möhnesee and Gibson's departure, which proved among the most memorable of the Second World War. The squadron commander was relieved to see no sign of searchlights, from which the dazzle could have been

Melvin Young (second from left) rowed in the victorious 1938 Oxford Boat Race crew. How Young earned the nickname 'Dinghy': here he is seen in a miraculous 8 October 1940 Atlantic rescue, the first of two following 'ditchings': (above) his crew in their dinghy after 22 hours adrift, spotted by a destroyer, and (below) with his rescuer's captain. (Right) With Priscilla Rawson, whom he married after a 1941 lightning proposal.

John 'Hoppy' Hopgood, whom Guy Gibson rated the squadron's best pilot; (right) with his sister Marna, an ATS officer, and (above right) the last family picture before Chastise, on leave with Marna and his mother. (Below) Aircrew of 106 Squadron, in which Hopgood served with Gibson, assembled beneath a Lancaster.

Henry Maudslay (above, second left), an Eton 'swell', member of 'Pop' and Captain of Boats just three years before he flew to the dams. (Right) The enormous strain of bomber operations had taken its toll, revealed by this 1943 image of 617's flight commander aged 21, taken for an escape kit picture, to be used if he was shot down.

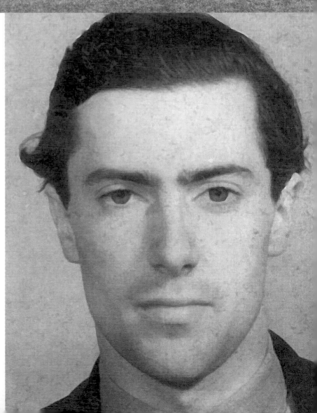

(Above) 617 Squadron's CO on the day of his investiture with Eve, his showgirl wife, then 30 to his 24. (Left) Gibson, photographed in 1944.

Guy Gibson as CO of 106 Squadron with the living creature he loved more than any other, Nigger, the slightly-mongrel black Labrador. Here he was supervising airfield vegetable gardening – and the dog.

Gibson with some of the crew who flew with him to the dams: (left to right) Fred Spafford, Bob Hutchison, George Deering, Terry Taerum.

(Top) Bill Astell on a family holiday in his pre-war life, and (above) as a bomber pilot in 1943; (above left) Gibson in in his office with David Maltby; (left) Five Australian members of one Lancaster crew photographed in London whilst on leave after the raid: (left to right) Leggo, Simpson, Hay, Foxlee, Martin.

(Above) 617's sole American, Joe McCarthy (third from right), with members of his crew; (far left) Canadian bomb-aimer John Fraser, wed to Doris a week before Chastise; (above left) Australian David Shannon; (below left) Les Munro.

The 133 aircrew who participated in Operation Chastise. Their names are listed on p.314

fatal to attackers making low-level bomb runs. Instead, there were only angry stabs of light flak, green, yellow and red, streaking up from six German positions, reflections on the water seeming to double their number.

At the outset, the gunners, a detachment from *Flakgruppe Dortmund*, were shooting wildly into the night sky, puzzled by the significance of the appearance of enemy aircraft, which could scarcely hope to breach the vast Möhne. Just before midnight, ahead of the warning from the Bismarck tower, a phone call from Luftwaffe regional headquarters at Schloss Schwansbell in Lünen alerted their crews to the proximity of enemy aircraft in the region. After months of boredom and inaction in this pleasant rural backwater, the gunners found it impossible to believe that the war had at last trespassed upon their haven, known complacently as 'the flak sanitorium' for those fortunate enough to have escaped, thus far, the multitude of unpleasantnesses imposed by global strife.

Of the three target dams, the Möhne offered by far the most open approach, from the south-east. To fly a heavy bomber head-on towards the automatic weapons on the walls nonetheless represented an immensely daunting challenge. Bomber crews hated light flak: 'Bit aggressive, aren't they?' said Gibson's Terry Taerum. While soldiers on battlefields can seldom see the bullets that may kill them, here the German 20mm tracer would be streaking towards each cockpit, every second of a crew's bomb run. Over the VHF link, Gibson ordered Hopgood and Martin to orbit the valley at the south end of the lake while he examined the approach: 'Stand by, chaps. I'm going to look the place over.' He reconnoitred the lake and dam from a distance that denied the gunners forewarning of their intended attack angle, then banked and climbed towards the other circling Lancasters, now joined by those of Young and Maltby, followed by Shannon. The last rashly strayed too close to the flak positions, perhaps three hundred yards from the nearest, and received two hits in the fuselage, prompting a protest from his bomb-aimer about pushing their luck. The hydraulic motors spinning the mines, which some crews heard running for the first time that night, were started several minutes before each made its approach. It was the wireless-operator's responsibility to monitor the rotation speed on a counter beside his set, twisting a knurled wheel to achieve a gradual increase in speed to 500 rpm, causing some aircraft to vibrate more fiercely than others.

It is a curiosity of what followed, that three-score fliers waged a private war above a masterpiece of *Kaiserreich* industrial construction, in which the only other participants were forty-odd German gunners, together with a few mesmerised, horrified civilian spectators. No fighters arrived; no generals or great men became involved. It was hardest for the attackers obliged to witness the travails of those who went before. Bomb-aimer Len Sumpter said: 'The worst time was when we were circling waiting to go in.' Once committed to an approach, each man was too preoccupied with his own task to spare much energy for fear. That recurred only when some glimpsed the imminence of their own ends.

Gibson told his crew, 'Well, boys, I suppose we'd better start the ball rolling.' Then he flicked on the VHF link, saying: 'Hello, all Cooler [the codeword for his wave] aircraft, I am going to attack. Stand by to come in to attack in your order when I tell you.' He added to Hopgood, his designated successor in command of the Möhne operation: 'Hello, M-Mother. Stand by to take over if anything happens.' 'OK, leader,' said Hoppy. 'Good luck.' Then Gibson dived his G-George towards the lake for the first operational trial of Barnes Wallis's Upkeep, the airframe vibrating as this extraordinary four-and-a-half-ton cylinder packed with explosive spun beneath the bomb bay. Since much about Bomber Command's fliers echoed the mood of students attempting some supreme athletic challenge, it does not seem frivolous to compare Gibson's predicament with that of a cricket captain. Under the eyes of his team, he was opening the bowling. It would betray everything that he represented if he let them down, muffed the moment, made a hash of his run.

It was 0028. As G-George lost height the first, most rewarding surprise for its crew was how clearly they could see everything under the brilliant moonlight – the dam wall, twin towers, sluices. Spafford called from the nose, 'Good show. This is wizard.' Then he suffered a moment's panic about the proximity of fir trees a few feet beneath his blister: 'You're going to hit them – You're going to hit those trees!' Gibson reassured him: 'That's all right, Spam. I'm just getting my height. Check height, Terry. Speed control, flight-engineer. All guns ready, gunners. Coming up, Spam.' Then came the vital forty-five seconds as they raced above the lake: navigator Taerum switched on the spotlights, saying, 'Down – down – down,' then, as the beams converged into a figure-of-eight, 'Steady – steady.' A red chinagraph mark on the air-speed indicator showed the 230 mph they sought, which Pulford made minute flap and throttle adjustments to hold. Spafford had long since fused Upkeep. Hutchison peered into the bomb's rotational-speed dial as front-gunner George Deering triggered his Brownings, though there is no record of any 617 Squadron gunner that night having inflicted injury on the Germans below.

Incoming tracer from the dam wall whipped past, some shells bouncing off the water: each man was braced at every

second for a brutal collision between German fire and the thin
alloy of the Lancaster. The bomb-aimer of another aircraft,
watching Gibson's run, said: 'We were perhaps a bit shattered
at the amount of flak – rather more than anyone had expected.'
Another bomb-aimer, Les Knight's Edward Johnson, thought
the gunners 'quite heroic, because it must have been quite
terrifying to be on the dam wall … when things started going
off'. Gibson again: 'This was a horrible moment; we were being
dragged along at four miles a minute, almost against our will,
towards the things we were going to destroy. I think at that
moment the boys did not want to go. I know I didn't want to
go. I thought to myself, "In another minute we shall be dead
– so what?"' He muttered to Pulford to be ready to pull him out
of the seat if he was hit – not that such action would save
anyone, at that height. The enemy fire was not nearly as heavy
as Gibson had experienced over German cities, but incompa-

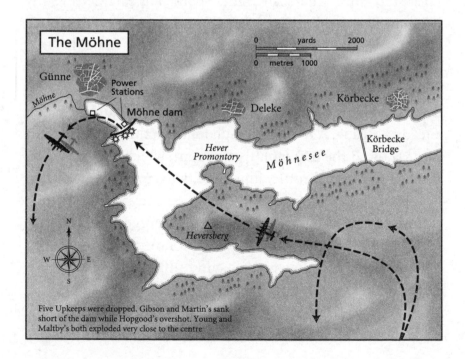

Five Upkeeps were dropped. Gibson and Martin's sank
short of the dam while Hopgood's overshot. Young and
Maltby's both exploded very close to the centre

rably closer. He felt almost overwhelmed by the vastness of the dam looming before his Lancaster, suddenly so small.

Deering's guns still hammered fire, his tracer glancing off the stonework of the southern flak tower, spent cases clattering from the Browning breeches. Directly beneath the gunner's feet, which rested on improvised canvas stirrups, Spafford peered through his Dann sight: 'Left – little more left. Steady – steady – steady – coming up.' Then: 'Mine's gone!' The aircraft surged sharply upwards with the loss of weight and Hutchison fired a red Very light, the agreed signal that their Upkeep was in flight. They hung low until, seconds later, after flashing over the dam flak positions, Gibson heaved back on the yoke and climbed away. Twenty-three-year-old local farmer's son Cpl. Karl Schütte, one of the gunners on the north tower, described how 'suddenly the speeding black shape was thundering like a four-engined monster between the two towers and over the wall at a height of about twenty metres, spitting fire and almost ramming the gun emplacement with its tail'. His gun's tracer hissed as it left the barrel, and he cried, 'Müller – keep shooting! Keep shooting!'

A voice over the radio called to Gibson from another aircraft: 'Good show, leader. Nice work.' The bomb bounced once … twice … three times – then vanished beneath the surface of the water. The approach run had lasted just forty-five seconds from the moment Taerum turned on the spotlights, yet it had seemed an eternity. Trevor-Roper was firing vengefully back at the dam gunners from the rear turret as a huge explosion sent a column of water shooting high into the night sky above the Möhne, drenching Schütte and his comrades, who spoke of 'a big black bird like a blurred shadow'. When Gibson's Upkeep exploded, a defender cried with shocked banality, *Mein liebe Herr*, they are dropping heavy ones!' The lake surface boiled, 'broken and furious, as though it were being lashed by a gale'. Dave Shannon saw 'a huge bloody spurt of water [that] went up hundreds of feet'.

Yet as minutes passed and the Möhnesee calmed, the dam still stood. Disappointment was crushing for Gibson – the leader, the best, the man always determined to show that he would ask no one to do what he could not do himself. Barnes Wallis may well have been correct, that just one of his bombs was capable of bursting the Möhne. But such a one needed to strike the dam with a precision that demanded more than most flesh and blood could contrive, or rather than most bomb-aimers could achieve with a Dann sight shaking in one hand amid the vibration of the rotating Upkeep, enemy tracer streaming towards them. The evidence suggests that Gibson's bomb was dropped early, causing it to sink before striking the dam wall. This reflected a fractional misjudgement by Spafford. After Gibson's extraordinary exertions, not least to overcome exhaustion and physical pain, his personal effort had failed.

It now became his task to rally and support the other attacking crews. The approach, he enthused over the VHF in an immortal RAF wartime phrase, wildly inappropriate here, was 'a piece of cake'. It was the turn of John Hopgood, who was obliged to continue orbiting for perhaps three minutes to allow the water to calm before he started his own run at 0033, just as Maudslay and Knight arrived to join the Lancasters circling above.

Survivors later testified that Hopgood and his front-gunner George Gregory had been wounded in M-Mother's earlier encounter with flak, though this could never be confirmed. All that is certain is that the Lancaster's attack went disastrously awry. German tracer raced to meet M-Mother the moment its spotlights were switched on. In the nose, John Fraser, the twenty-year-old British Columbian bomb-aimer who had already completed a full tour of bomber operations, peered through the eyehole of his wooden sight, then felt cannon shells thrash the aircraft, causing an instant eruption of flames. 'It's burning! It's burning!' cried exultant gunner Karl Schütte on the dam wall, even as his crew kept firing. A voice in Gibson's aircraft exclaimed, 'Hell! He's been hit.' Hopgood, in the stricken plane, called to Fraser to drop the Upkeep, which struck the water even as the burning Lancaster cleared the dam, struggling to gain height. The bomb sprang over the wall and hurtled two hundred feet downwards, to explode on the Möhne's power station. Blast caused fragments of masonry to rattle and thud down around the gun position, hurling Schütte to the ground, and obliterated the electricity installation. A red flare soared from Hopgood's aircraft, the signal that his bomb had gone, but none of the watchers doubted its irony, because they saw that the Lancaster was doomed.

Hopgood somehow clawed five hundred feet in a banking turn to clear the forested ridge north of the Möhne, then, less than thirty seconds after breasting the dam wall called to his crew, 'For Christ's sake, get out of here!' It was a significant scandal of the bomber offensive, for which Avro shared responsibility with the RAF, that parachute escape from a stricken Lancaster was exceptionally difficult because of its inadequate emergency hatches. Bomber Command's Operational Section repeatedly highlighted this issue: whereas half the fliers aboard doomed USAAF bombers survived, only one in five of the RAF's did so, and just 15 per cent of Lancaster crews. Yet nothing was done. In those terrible seconds aboard the blazing M-Mother, only three of Hopgood's crew were able to make escapes. Bomb-aimer John Fraser ditched the emergency hatch immediately beneath his position. Seeing trees grotesquely close, before jumping he burst open his parachute pack, thrusting it before him as he jumped. The canopy had just sufficient time to deploy before Fraser hit the ground, by some freak of fortune unhurt, but traumatised.

Tony Burcher, the Australian tail-gunner, was by all accounts an unruly character, once sentenced by Gibson to a 'dry' stint after fighting in the mess. He later described how he pushed out of the aircraft M-Mother's badly wounded wireless-operator, Gloucestershire baker's son John Minchin, pulling the ripcord of the man's parachute before opening his own canopy and then jumping. Whether or not this is accurate, Minchin failed to survive: his body was found some distance back from the place, three miles north-west of the dam, where M-Mother ploughed into the ground and dissolved into flames, killing the four men still aboard. Just six weeks later, Minchin's younger brother was also killed flying with Bomber Command. Burcher meanwhile landed alive, a mile or so back towards the dam, but with a badly injured back. After hiding in a culvert for two days, he was taken prisoner.

'Poor old Hoppy,' said an unidentified voice among the crews still circling the target, stunned by the horror of the spectacle they had witnessed, and knowing the significance of the flash of light that briefly illuminated the sky beyond the hill. Only three minutes had passed since Hopgood began his attack. M-Mother's Upkeep was released far too late; the post-war testimony of bomb-aimer John Fraser should be treated with caution, but there must have been confusion in the aircraft after it was hit.

So much has been written about the courage of the aircrew who attacked the Möhne that some words also seem deserved about that of the gunners who defended it. The fliers enjoyed at least the advantage of understanding what they were attempting to do, supremely hazardous though it was. The Germans, by contrast, found themselves faced by a wholly unknown phenomenon, elemental and mortally threatening. Hitler did not dispatch his brightest and best, SS supermen, to guard a hitherto inviolate target in the depths of the German countryside. Yet the Möhne's men fought their guns with a conviction worthy of a better cause. After destroying M-Mother, with scorched fingers they changed the barrels of their superheated guns, reloaded, awaited the next round. The good news for the attackers, had they but known it, was that the crew on the south tower reported now their weapon jammed and unserviceable – a failure such as is often suffered by intensely-fired automatic weapons.

A few hundred feet above, at 0037 Hutchison sent a terse Morse message to 5 Group – the codeword 'GONER68A', reporting the dropping of G-George's bomb. The '6' signified that it had landed within five yards of the dam, probably an overstatement; the '8' indicated that it had not achieved a breach at 'A' – the Möhne. In Cochrane's underground operations room at Grantham, Bomber Command's C-in-C had joined Cochrane and Barnes Wallis. The presence of so much brass surely intensified the engineer's gloom and chagrin.

At the Möhne, it was the turn of Micky Martin. In the unpublished draft of *Enemy Coast Ahead* Gibson described the Australian equivocally as 'slightly more ham' than some other pilots. This was later changed to the more flattering 'splitarse', to which the squadron commander added, 'Flying to him was nothing unless it was dangerous.' It is not difficult to divine that the disciplinarian Gibson was sometimes irked by this wild Antipodean, though acknowledging him as an outstanding flier.

Gibson called over the VHF: 'Come in number three, you can go in now.' He had made a decision that, though his crew never afterwards complained of it, may have caused them some anguish. Having braved enemy fire to reconnoitre the dam, then made his own attack, the squadron commander now proposed to fly alongside Micky Martin as he dropped his bomb, to divert German fire. He would invite his own destruction, the fate that had just befallen John Hopgood and his men, to improve another crew's chance of success. Gibson had been an exhausted man even before he took off. Only iron strength of will could be keeping him airborne.

Battlefield heroes are seldom popular among their own comrades. Officers who court danger are especially prone to incur the dismay and even anger of those under their command, who mutter, 'It's all right for *him*, if he wants to win a VC. But what about *us*?' The majority of those who fight, including men such as those who formed 617 Squadron, are committed to doing their duty, but yearn to survive to go home. It would be strange if, in G-George at 0038 on 17 May 1943, some of Gibson's crew did not nurse harsh thoughts about their skipper, as once again their Lancaster turned and lost height to approach the dam, just ahead and to starboard of P-Popsie.

Martin was a brilliant airman, as cool and courageous as Gibson. His bomb-aimer Bob Hay was bombing leader for the whole squadron, and thus its anointed expert at his craft. As the Lancasters roared up the lake, both front-gunners firing continuously, the returning stream of German cannon shells targeted the bomb-dropping aircraft, not Gibson's. Smoke was still curling from the shattered power station, drifting over the south tower. Hay was a millisecond from pressing his release when P-Popsie was hit by shells in the starboard wing. The strikes seemed to jolt the aircraft just as its Upkeep fell away – and veered left. Hay afterwards asserted his conviction that the drop had been perfect. Be that as it may, the Upkeep bounced across the water, finally exploding short of the dam wall, a critical fifty yards off the central aiming point. The cannon strikes seem to have shaken the plane or distracted its crew at just the moment when precision was indispensable. As Martin, too, banked away, his rear-gunner saw a gigantic waterspout shoot skyward as the Upkeep exploded – and left the dam standing.

The three outstanding fliers of 617 Squadron had now failed in attempts to breach the Möhne, prompting deepening dismay as successive 'GONER' signals were received at Grantham. In a later letter to Cochrane, Wallis looked back on those terrible minutes in the 5 Group operations room, following news that a succession of Upkeeps had achieved nothing. 'You will understand, I think, the tremendous strain which I have felt at having been the cause of sending these crews on so perilous a mission … the tense moments … when I felt that I had failed to make good were almost more than I could bear.'

Dave Shannon later admitted: 'There was a certain amount of disappointment … we'd been told that one of these bombs placed in the right position would blow up the wall.' The overhead aircraft were once more obliged to linger, circling above the Möhne for four minutes to allow the lake to calm before, at 0043, the turn of Melvin Young came. This time Gibson overflew the reservoir on its northern side, flicking his identification lights on and off as an additional distraction for the flak. Once again he was inviting his own death. Some of those in G-George assuredly mutely cursed Barnes Wallis, Ralph Cochrane, and their own impotence in the hands of a skipper prepared to embrace any course, to accept any sacrifice, to fulfil the task that had been committed to him.

Martin meanwhile took station on Young's port wing, matching his approach, while the gunners of all three aircraft kept firing. A-Apple flew low, low, lower up the lake, its navigator Charlie Roberts calling the now familiar 'Down – down – down – steady' while bomb-aimer Vince MacCausland peered through the eyehole of his Dann sight. Then their load was gone, skipping over the water while the two bombers banked to starboard and climbed steeply away. Upkeep bounced three times; struck the dam wall; vanished. The positioning was perfect. Three seconds later the hydrostatic pistols fired the charge. Six thousand six hundred pounds of Torpex exploded, inflicting upon the Möhne a pulverising earthquake shock. As a new column of water soared upwards, Young cried exultantly over the VHF, 'I think I've done it!' Shannon, circling above, thought, 'Christ, that must break the bloody thing.' Yet as the lake calmed and the water subsided, still the wall appeared unbroken. Micky Martin cried into the VHF, 'Wizard show, Melvin. I think it will go on the next one.'

At 0049 it was the turn of David Maltby, who had celebrated his twenty-third birthday a week earlier, and had a wife back in Lincolnshire expecting a baby in July. Among 617's most experienced pilots, he made another perfect run, perhaps assisted by silence from Karl Schütte's gun on the north tower, now also jammed after thirty minutes of almost continuous firing, though Maltby's navigator noted in his log: 'Flak none too light.' Some of the despairing Germans below were reduced to firing their rifles towards J-Johnny. 'It seemed almost laughable,' said Schütte, 'but it took our minds off the danger we were in.' Then Maltby's Upkeep was in flight – and even as the aircraft crossed the dam wall, the pilot saw its crown already crumbling. Edward Johnson, watching from N-Nuts, said that in the moonlight 'We saw the bombs bouncing quite clearly.' Maltby's Upkeep, too, struck the dam and exploded, sending water – this time accompanied by mud and fragments of masonry – soaring into the sky. As the pilot banked he saw the eruption silhouetted against the moon: '[The spout] rose with tremendous speed and then gently fell back' to reveal a colossal breach opening in the midst of the dam. Gibson banked towards the wall, and saw the lake surface 'looking like stirred porridge … gushing out and rolling into the Ruhr Valley towards the industrial centres of Germany's Third Reich'.

There was a thrilled yell over the VHF: 'It's gone! It's gone!' With firing from below now much weakened, a succession of Lancasters dipped to view the torrent they had unleashed. The desperation that had overtaken the fliers, the sense of humiliating failure, was replaced by exultation, exhilaration, wonder, relief, triumph. In every watching cockpit there were cheers and roars across the intercoms. 'I can hardly describe the atmosphere in the plane,' said Maltby. 'The yelling, the pure excitement.' These very young men succumbed to a schoolboy joy. There has since been debate about whose bomb breached the dam. It seems almost certain that Young's had achieved a decisive fracture, which became apparent even as Maltby's exploded. Gibson would obviously have aborted J-Johnny's run had the breach been certain before its bomb-aimer pressed the release-toggle.

Now the squadron commander watched the tidal wave sweeping down the valley, the clouds of spray that became a fog, the irresistible weight of water racing through the breach in the dam, and ordered Hutchison to send to Grantham the historic one-word success code: 'NIGGER'. When the signal reached Cochrane's operations room at 0056, the receiving staff officer demanded repetition, confirmation.

In the preceding ten minutes Harris, Cochrane, Wallis and their staffs at St Vincent's had received a dispiriting succession of messages. Following Gibson's 'GONER', at 0050 the same signal came from Young's aircraft, then belatedly at 0053 'GONER58A' from P-Popsie, indicating that Martin's mine had exploded fifty yards from the aiming point. These tidings caused Barnes Wallis to bury his head in his hands. He had overcome huge bureaucratic, scientific and technological obstacles; had caused industrial and technical resources to be poured into fulfilling and now testing in action his creation, at the cost of lives. The fliers, the young men, were doing their parts with wondrous courage. The old ones who had sent them, however, Wallis foremost among them, appeared to have got it wretchedly, tragically wrong.

And now, the confirmatory codeword: 'NIGGER', repeated by Hutchinson. In Wallis's long life, much had gone before, and more was yet to come. Yet this seemed his finest hour, the success that not merely vindicated his struggle to build a weapon that would smash Germany's dams, but also secured his place in history, and in the British people's narrative of the Second World War. He jumped up, pumping his arms like a triumphant athlete. Every face in the operations room broke into beaming of relief and congratulation. Gibson later described Wallis as 'the man who was a witness of his own greatest scientific experiment in Damology'. Harris, whose bleak countenance masked an even bleaker view of warfare and of mankind, shook the 'boffin's' hand. 'Wallis,' he said, in phrases that became famous, 'I didn't believe a word you said when you came to see me. But now you could sell me a pink elephant.'

At the Möhne, one gun was still firing from the dam wall, and others from below, as the circling Lancasters surveyed the astounding torrent creating a mist that now began to overhang the woods below the dam. Gibson wrote: 'Down in the foggy valley we saw cars speeding along the roads in front of the great wave of water which was chasing them and going faster than they could ever hope to. I saw their headlights burning and … water overtake them, then the colour of the headlights underneath the water changing from light blue to green, from green to dark purple, until there was no longer anything except the water, bouncing down in great waves.'

Gibson now sent home Martin and Maltby, whose Upkeeps were gone. He himself, still leading the way, turned towards the Eder dam, forty-five miles eastwards over the hills and valleys of the rustic Sauerland, with the three First Wave crews whose Lancasters still carried bombs – those of Shannon, Maudslay and Knight. They were accompanied by Melvin Young, who was mandated to assume direction of the Eder attack if the CO went down. This was an odd decision by Gibson, which can only have reflected lack of confidence in Maudslay. Young, who had expended his Upkeep to break the Möhne, was required to linger impotent for an additional hour over Germany. There is no record of what the Eder-bound crews thought during the ensuing thirty minutes. The thrill of success at the Möhne was overlaid by apprehension about what they themselves still had to do, together with awareness that the Germans had now seen their dams under deadly attack, and still had time to dispatch fighters to fall upon the Lancasters.

At exactly the moment that Gibson's First Wave began to assault the Möhne, across the hills ten miles to the south, 617 Squadron's American pilot and his crew were acting out a matching drama, the more moving because of its loneliness. Five Lancasters had been briefed to bomb the Sorpe, yet just one of them survived to do so. Joe McCarthy reached the dam at fifteen minutes past midnight, and set about studying the formidable challenge it represented. The Sorpe was an earthen

construction, 2,250 feet in breadth, set in a valley densely
wooded on the eastern side, while on the rising western face
stood the village of Langscheid. The river that ran below it
joined the Ruhr just two miles below the dam wall. Unlike the
Möhne and the Eder, unmistakably man-made constructions,
the Sorpe almost resembled a natural feature. The dam was
undefended, so the only sound to break the tranquillity of the
area that night was the roar of the Lancaster's Merlins as
McCarthy and his crew explored alternative approaches,
groped for the best means to place their mine alongside the
wall, their task made more difficult by the absence of towers
such as provided guiding marks for Gibson's Wave.

The Air Staff had written succinctly about the Sorpe and
Wallis's bombs: 'worth destroying. But its construction is such
that great difficulty would be experienced in destroying it by
the means at present under consideration.' Because of the steep
slope on the lake side of the dam wall, the Sorpe was invulnerable to a bouncing waterborne delivery. The planners thus
ordered the attacking crews to make a lateral approach, flying
along the wall from the west without activating their coun-
ter-rotational drive, then releasing their Upkeeps so they sank
alongside. Wallis never supposed that a single bomb could
thus blow a breach. He nonetheless hoped that a succession of
giant explosions might do so.

After the event, Arthur Collins of the Road Research
Laboratory, who had unlocked the hydrological and ballistic
keys to breaching the Möhne, suggested that if he had been
invited to study the Sorpe's structure and to conduct experiments ahead of the attack, he might have found an answer. In
truth, it is unlikely that any number of the weapons available
to the wartime RAF could have broken the Sorpe – in 1944,
deep-penetration twelve-thousand-pound 'Tallboys', created

by Barnes Wallis, would fail to do so. Moreover, if all of 617's crews were invited to do the near-impossible on the night of 16/17 May, those dispatched to the Sorpe faced special difficulties. They were required to make a sharp descent from the village side to the lake; during a seven-second passage along the wall they must release a bomb at a briefed airspeed of 180 mph, with a margin of error of only a few feet, to sink close enough for detonation to precipitate the necessary earthquake effect; then climb away before crashing headlong into the steep, pine-studded hillside beyond the dam.

It was impressive that Joe McCarthy and his crew found their way to the Sorpe so quickly, after their delayed take-off. He and the crew of T-Tommy thereafter spent thirty-one minutes manoeuvring around a small patch of night sky in the heart of Germany, making successive attempts to achieve a run that satisfied Johnny Johnson, their bomb-aimer. The inhabitants of Langscheid were bewildered as well as terrified – how could they not have been? Some had already glimpsed tracer arching into the sky around the Möhne, beyond the hills. Now came the roar of a huge aircraft thundering impossibly low over the rooftops of their homes, again and again brushing the steeple of their church, apparently to no purpose.

The Sorpe was defended by a single elderly Home Guard, whose job was to watch for saboteurs, not Lancasters. Meanwhile machinist Josef Ketting was asleep in his quarters at the power station below when he was shaken awake by his wife, alarmed by McCarthy's first low pass. Having dressed and gone outside, Ketting met three workmates who had received telephone warnings of bombers in the region. He was just considering returning to bed when T-Tommy again zoomed overhead, so low that he could see its RAF roundel. He told his wife quickly to fetch their son and family valuables, then seek

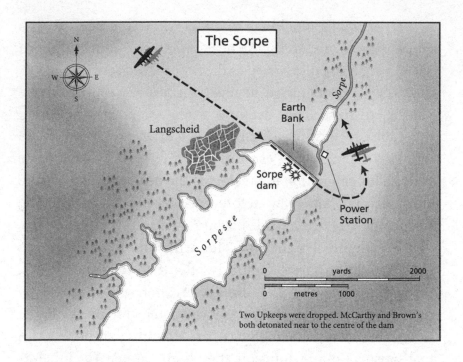

Two Upkeeps were dropped. McCarthy and Brown's both detonated near to the centre of the dam

shelter in the cellar of the nearby Brinkschulte inn. It did not cross the Germans' minds that the British might be so ambitious as to aspire to break the dam. They supposed that, instead, their objective might be to destroy the power station below.

The persistence and dedication with which T-Tommy's crew strove to fulfil their mission received much less subsequent applause than did the heroics of Gibson's First Wave, because McCarthy and his men failed. But their attempt was at least equally deserving of admiration. T-Tommy made nine aborted approaches across the Sorpe, without dropping its Upkeep. Again and again, Johnny Johnson called 'Dummy run!' before they pulled away. As the minutes ticked by, the exasperation of his crew rose. 'Won't somebody get that bomb out of here?' demanded a voice on the intercom – surely with an obscene adjective attached. The Germans below must by now have alerted Luftwaffe headquarters. At any moment, enemy fighters could appear. The longer they lingered, the more alert must be the ground defences on their route homewards.

The record shows that some crews of Bomber Command's Main Force released their bombs at the first plausible opportunity on an approach to a city. Pilots who circled twice, to ensure accuracy – as Gibson had sometimes done in his earlier incarnation – were often cursed by their comrades as lunatics. Yet here now was T-Tommy, its pilot and bomb-aimer stubbornly committed to ensuring that Wallis's weapon was released where it might work. McCarthy struggled to find a way of clearing the village, then dropping low enough, fast enough, to make his run. 'I had trouble with that darn [church] steeple,' he said later. Their eventual tactic was to align the port outer engine with the dam wall. In the midst of a tenth run, suddenly the plane lurched and Johnson called, 'Bomb gone!'

More than four tons of Upkeep sailed downwards, sinking astonishingly close to the dam – barely twenty feet from the crown of the wall and forty yards south of its centre point. Then came the vast explosion of T-Tommy's weapon, from which debris showered the crest of the wall. Johnny Johnson said: 'When the mine exploded it was a terrific sight. We were nose up at that stage and turning. The explosion was between our aircraft and the moon and Dave Rodger in the rear turret had a clear view' – indeed, water spouted up around the gunner. To Josef Ketting, watching below, the object that he saw fall from the Lancaster's belly resembled an oversized farm slurry tank. His first thought after the explosion was that when daylight came, there would be a rich harvest of fish to be netted in the lake.

T-Tommy orbited once more, to explore what they had achieved: 'We saw it had crumbled about two-thirds of the way along the dam wall. We were hopeful – but that was as far as it got.' For a few ecstatic moments, McCarthy believed that they had breached the dam, until navigator Don MacLean forcefully urged him to focus on flying the plane – keeping them alive. The crew sent their 'GONER79C' message – signifying failure – to 5 Group only when they were twenty minutes out from Scampton on the way home; it is unknown whether this lapse derived from forgetfulness or radio-transmission difficulties. The Sorpe stood, as it continues to stand seventy-six years on, though after the war the reservoir was drained to repair injuries inflicted by the RAF's two wartime attacks. McCarthy and his crew had nonetheless accomplished all and more than their commanders had sought from them. The pilot banked to port, climbed, set course for home. A few minutes later they passed over the raging Möhne valley. Though their own effort had failed, they saw that those of others had succeeded. They marvelled at the gaping breach in the dam, the ever-widening, lengthening flood in the valley below, then turned to retrace their outbound course towards Scampton.

2 THE EDER AND AFTER

Mist was beginning to creep into the valleys as Gibson approached the western end of the Edersee at 0130, after a sound piece of navigation by Taerum. He flew low, to examine the lake's snaking waterways, inlets and promontories before banking over the dam. Only Dinghy Young arrived as promptly. None of the other three aircraft was in sight when Gibson broke radio silence to call over the VHF: 'Hello, Cooler aircraft – can you see the target?' Shannon, in L-Leather, responded that he thought he must be nearby. Gibson said he would fire a red flare over the dam, which the Australian promptly spotted, and realised that he himself was above the wrong end of the lake. Within minutes, all five Lancasters were flying circuits over the hills, studying the daunting problem of a low-level approach. No model of the Eder had been available at briefing. 'Nobody had seen it before,' said Len Sumpter, Shannon's bomb-aimer. 'They didn't know what the situation was.'

The reactions of flak crews across Germany to the passages of 617's aircraft that night were impressively lethal, but it remains an enigma why the Luftwaffe failed to scramble fighters to any effect. To be sure, neither ground-based nor airborne radar was of any value against such low-flying attackers, but by the time Gibson and his followers approached the Eder, the region's dams had already been under assault for seventy-five minutes, an eternity in the timelines of air battles, when the defenders possessed relatively sophisticated telephone and radio networks. Much of what was done that night reflected the RAF's skills and courage, but crews also had a significant measure of luck. The Eder had no flak defences – its few guns had been removed. Gibson and his thirty-four men were able to manoeuvre above it for the next twenty-four minutes, the darkness unbroken by enemy activity either in the air or on the ground.

Yet this did little to diminish the difficulty of their task. Indeed, the natural hazards fortifying the Eder made it the most terrifying of the night's objectives. On all sides, hills rose two or three hundred feet. A jutting, densely wooded promontory lay in the midst of its lake, less than a mile from the centre of the dam. This enforced a dogleg approach, leaving pitifully little time in which to align a crew's bomb run. And as a pilot flew up a natural cul-de-sac towards the aiming point, it was hard for him not to peer beyond the dam at the Michelskopf, a sheer wall of wooded mountain a mere mile beyond, which confronted an approaching aircraft. It was possible to release a bomb and live to tell the tale only by making an almost instantaneous violent starboard climbing turn, to gain a thousand feet. This would be difficult and dangerous enough in a nimble single-seat fighter weighing three or four tons. To do so in a thirty-ton heavy bomber was akin to inviting an elephant to emulate a gazelle. On the port side of the approach, less than two miles back from the dam, towered the turreted walls of Schloss Waldeck, four hundred feet above the water, commanding magnificent views. Anyone who stood upon the castle's battlements and considered the challenge facing Gibson's men would have declared that the commanders who ordered fliers to execute such a mission were callous fantasists. The principal reason the dam was undefended was that the Germans believed nature provided protection enough.

Though Gibson and his crews explored the Eder for twenty minutes before the first Upkeep was released towards it, each pilot found the approach terrifyingly brief. From the moment a Lancaster breasted the densely wooded ridge at the northern end of the lake, dropped down to the water for the bomb run, released its bomb, crossed the dam and banked desperately in pursuit of clear sky, no more than forty-five seconds elapsed – but there was so much to be done in the time. Len Sumpter said: 'By the time you had got the [correct] height, the towers were through the sights.'

Shannon attacked first. Most of 617's aircraft bore no personal identification marks, because crews had owned them so briefly. L-Leather nonetheless was adorned by the hastily-painted name 'Bacchus', because its crew boasted so many serious drinkers. The twenty-year-old pilot, son of a South Australian farmer and state assemblyman, was dating Scampton WAAF officer Anne Fowler. At that moment, the prospects for their future relationship seemed precarious in the extreme. Shannon made a left-handed approach, turning as sharply as he dared after passing beside Schloss Waldeck, then dropping steeply. When he levelled out for his attack 450 yards short of the dam, accusing fingers – the beams of the spotlights below his fuselage – told him that he was still high – too high to release his bomb. The engines roared at maximum power as he flashed over the dam wall, then swung to starboard and climbed as hard as the huge, still-laden plane allowed, to escape the four-hundred-foot rock face ahead. In a heavy bomber, everything happened in slow motion. 'To exit from the Eder Dam with a 9,000-lb mine revolving at 500 revs was pretty hairy,' said the pilot later. Though ten days short of his twenty-first birthday he enjoyed a reputation as an outstanding flier, but 'it was a bugger of a job'. Safely above the hills, after giving himself and his crew an epochal fright, he made an anti-clockwise turn back down the lake for a second attempt. Once again he descended as fast as he dared; once again he managed to steer a correct course – but was still too high. The same happened on a third, then a fourth run: each time the pilot shouted to the bomb-aimer to abort. Gibson called, 'OK Dave, you hang around for a bit and I'll get another aircraft to have a crack.' With just three Upkeeps left, it was vital not to waste one.

Henry Maudslay's Z-Zebra was next to take its turn. The crew started the little engine to rotate its huge bomb, then as the aircraft began to judder from the movement of driving wheel, V-belt and Upkeep, they braced themselves in their seats. 'B' Flight's commander cannot have been in a happy state of mind. Beyond the apprehension inseparable from what he was charged to do, Gibson had made explicit a lack of confidence in him, by appointing Hopgood and Young deputy leaders for the two dams' attacks. This decision was probably influenced by Maudslay's performance in training, and especially by his accident at Reculver.

Now the pilot was called upon to make the fierce effort of concentration essential to survive an approach up the Eder lake. The problem was the same as that faced by Shannon: to lose height fast down to sixty feet, after clearing the wooded promontory. Maudslay dived, levelled – then found himself in the same bind as the Australian: too high. There are conflicting accounts of some events at the Eder. Les Knight's front-gunner, Fred Sutherland, later reported Gibson rebuking Maudslay over the VHF after this first run, saying, 'Henry, that's very nice flying, but you will have to do better than that,' to which the pilot replied, 'Sorry, sir, I will try again.'

On Maudslay's second circuit, the same happened. Gibson ordered him to circle, while at 0139 Shannon renewed his attack – and yet again failed to make the drop. Every man

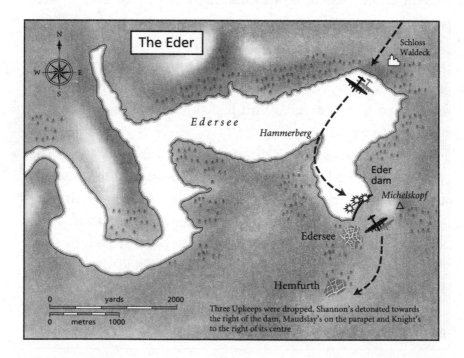

Three Upkeeps were dropped. Shannon's detonated towards the right of the dam, Maudslay's on the parapet and Knight's to the right of its centre

knew that to release an Upkeep on the wrong course, at the wrong height, represented a squandering of life and effort. At the Möhne, they had been shown that absolute precision was essential to make Wallis's weapon work as designed. Only on Shannon's sixth run was Len Sumpter, in the nose, convinced of success. He stabbed the toggle, the bomb fell away, bounced twice on the water, struck the dam and exploded, thrusting into the sky the now familiar fount of water, a few yards on the shoreward side of its right-hand tower. As the Lancaster climbed and the lake settled, Shannon believed, and continued to assert until the end of his days, that his bomb had achieved a small breach, but there is no evidence of this. Gibson wrote: 'Dave made a good dummy run, and managed to put his [bomb] up against the wall, more or less in the middle. He turned on his landing light as he pulled away, and we saw the spot of light climbing steeply over the mountain as he jerked his great Lancaster almost vertically over the top. Behind me there was that explosion which, by now, we had got used to – but the wall of the Eder Dam did not move.'

At 0145 Maudslay approached again – dropping, dropping, dropping until his bomb-aimer Michael Fuller, a former tele-phone engineer from Reigate in Surrey, pressed the toggle. Z-Zebra's Upkeep was gone ... but too late for even a single bounce on the water. It flew onward through the air to the dam, to detonate on the parapet. Shannon said: 'It was the same as with Hopgood: his bomb hit the top of the wall, bounced over.' As the valley was brilliantly lit by the lightning bolt of the explosion, Gibson glimpsed the Lancaster caught in its blast. 'It all seemed so sudden and vicious,' he wrote in his memoir, 'and the flame seemed so very cruel.' A red signal curled skywards as Maudslay's wireless-operator fired his flare pistol from the mounting behind the astrodome in token of the dropping of Upkeep, but every man in the four circling Lancasters was convinced that the aircraft was doomed.

'Henry, Henry!' Gibson called urgently over the VHF. 'Z-Zebra. Are you OK?' Hearing only silence, he called again, and heard a faint response: 'I think so – stand by.' Maudslay's voice on the R/T sounded to Gibson dazed and unnatural; to another flier 'almost dehumanized'.

This was their last contact with the crew of Z-Zebra. Though the others saw no further explosion, they assumed that the aircraft was lost. They had no knowledge that Maudslay's Canadian wireless-operator Alden Cottam transmitted the message acknowledging failure – 'GONER28B' – to 5 Group eleven minutes later, at 0157, about the time Z-Zebra would have been passing the Möhne on what was later discovered to have been a limping passage homeward. Some damage to the Lancaster at the dam must be assumed from the fact that its pilot made no further VHF voice contact with Gibson. On what had proved to be Maudslay's final run, there is no doubt that his Upkeep was dropped too late, and off course. Uncertainty persists, however, about whether the aircraft was flying too fast, or had suffered earlier damage – if so, curiously unreported to Gibson – en route to the target, or merely fell victim to the fractional misjudgement of pilot or bomb-aimer.

At 0142, and then again three minutes later, Gibson radioed Bill Astell, in hopes of summoning a reinforcement to the Eder. He heard nothing, however: Astell's shattered corpse was already cold. Just one Upkeep remained. Les Knight in N-Nuts, circling the end of the lake, waited for smoke from the explosion of Maudslay's bomb to clear from the dam. He was a former trainee accountant from the Melbourne suburbs, untroubled by comrades' teasing about his sober habits. An

immensely conscientious twenty-two-year-old, he had
survived twenty-three operations with 50 Squadron.

617's adjutant, Harry Humphries, characterised Knight as a
'pleasant-faced boy, [who] used to write reams of letters home
to his native land. His idea of a hectic night out was a visit to
the cinema. He had no [love] affairs as far as I know, merely a
burning desire to return home when his job was done. He
talked a lot about sheep-farming, used to fill in a lot of appli-
cations for jobs in Australia.' Like the others in N-Nuts, Harry
O'Brien, the Canadian rear-gunner, felt warm admiration for
the pilot with whom they had already shared so much – by no
means a universal emotion among crews towards their skip-
pers: 'He was the coolest and quickest-thinking person I have
ever met. And, in my opinion, the most knowledgeable in the
squadron with respect to his job.'

Knight made his first run at 0147, with other crews overhead offering radio advice so freely that the exasperated pilot,
desperate to concentrate, switched off the VHF link. Unhappy
about his course and height, he aborted without dropping the
bomb. He had learned something, however, and swung around
the lake for a second attempt convinced that he must make an
approach to the left of the track chosen by Shannon and
Maudslay. Ray Grayston, Knight's flight-engineer, also
concluded that the best way to achieve the right height and
speed was to drop fast to sixty feet by cutting back the engines
almost to idle as they dipped towards the water, then thrust the
throttles through the gate, praying that they would regain
power instantly. The great Merlins did not fail them: at 0152,
they drove N-Nuts pell-mell down the lake.

Sid Hobday, Knight's navigator, said: 'As we came in I wasn't
tense. I was excited. It was a great thrill if you've been on a
bombing squadron as long as we had … On this occasion it

was marvellous.' Edward Johnson, in the nose, felt differently: 'It's quite terrifying to see [the dam] looming up at high speed.' Seven seconds later the bomb-aimer made a perfect drop at 240 mph, 450 yards from the wall. This represented a supreme team effort by pilot, navigator, flight-engineer, wireless-operator and bomb-aimer. It helped that Knight's crew had been together so long; they knew each other more intimately than any other in the squadron. Grayston, a twenty-four-year-old former ground-crew mechanic from Surrey, said, 'Then we had to climb like buggery.'

Upkeep bounced three times, hit the wall, sank. There was the now familiar huge explosion; a pillar of rising water, together with a brilliant flash, prompted by a short-circuit in the power station below. Knight's wireless-operator, Robert Kellow, peered back at the dam: 'It was still intact for a short while, then as if some huge fist had been jabbed at the wall, a large, almost round black hole appeared and water gushed as if from a large hose.' Edward Johnson, in the nose, missed seeing the collapse, but knew about it quickly enough from the exultant cries of their rear-gunner. 'We forgot all about safety and going home and we were trying to follow the water down the river … It was a terrifying sight. We could see cars being engulfed.' Harry O'Brien in the rear turret suddenly felt 'a great joy. We felt the exquisite pleasure one feels when he had completed a difficult task perfectly.'

A surging torrent swept down the valley, across tens of thousands of acres of north-west Germany. 'We could see the water gushing out,' said Sid Hobday, 'and all the masonry coming down. It really was fantastic. It's a sight I shall never forget.' In the nose of Shannon's L-Leather, Len Sumpter saw the breach widening: 'the water was washing the sides out and also the wall below'.

Werner Salz worked as a fitter in the Eder power station. Earlier that Sunday evening he had taken a stroll with friends, on which they met some people who had attended an air-raid precautions meeting in nearby Hemfurth. They said it had been explained to them that they should have no fears for the dam, which was invulnerable to air attack, even in the unlikely event that such a thing came to pass. The local air-raid precautions chief had warned sternly: 'Anyone who leaves the shelters during an alert is a traitor to the Fatherland.' Salz was awoken by the coming of the Lancasters. He dressed and climbed a nearby hill, from which he could see the dam. Each explosion caused a tremor in the ground under his feet. The third detonation prompted a distinct judder, which the young man immediately interpreted as a shock to the dam. He ran back to the hamlet, to find homes already encircled by water, some people fleeing for their lives, others searching for families. Then he thought of the four men working the night shift in the power station, and raced to the dam wall to behold 'an enormous hole, thundering masses of water, a catastrophe'. Three workers were fortunate enough to be able to clamber to a refuge on the roof of the switch room, but one drowned below, and his corpse was discovered only three months later, when the turbine exit channel was drained.

On the morning of 16 May, the Eder lake contained seven hundred million cubic feet of water. Seven hours after the dam was breached, the level had fallen by over thirty feet: more than half the contents of the Edersee had poured through the breach, to be followed by a further two hundred million cubic feet in the ensuing two days.

At 0154 Gibson dispatched to Grantham the success code-word 'DINGHY', prompting a resurgence of the euphoria that had overtaken Cochrane's operations room. This remained undampened when 5 Group received Gibson's response to its question of whether his Wave still carried any Upkeeps that might now be directed to Target Z – the Sorpe. There were none, Cochrane's staff was told. Harris, the former extravagant sceptic, now seized proud ownership and fatherhood of Chastise. He announced that he would telephone news of its triumph to the chief of air staff in Washington. There were difficulties in persuading successive service switchboards to connect the C-in-C to the White House, but these were eventually overcome. In America's capital, it was a spring evening. Harris conveyed the brilliant tidings to Portal, prime mover for the operation. The CAS, in turn, promised to inform the prime minister immediately.

In the course of the Second World War, it was often the task of British service chiefs, and of Churchill himself, to explain to the Americans with varying degrees of embarrassment, and occasional shame, why they could not do some things; had failed in attempts to do others; lacked resources to accomplish more. It was wonderfully cheering for Britain's leader now to be able to report to America's president and warlords news of a brilliant success, a marriage of British imagination, technical brilliance and courage.

At the Eder, Gibson called up his crews and said, 'Well, it's all right boys, you're having a good time – but we've still got to get back to base. Let's go.' Another pilot remembered slightly different phrases: 'Good show, boys, let's all go home and get pie' – though the last word may in truth have been more vulgar.

Yet if the night's principal events were over for some crews of 617 Squadron, for others they had scarcely begun. Ken Brown's F-Freddie completed its North Sea crossing at 0130, just as Gibson's men reached the Eder. The planners intended that the five aircraft of the Reserve Wave should respond to circumstances and opportunities – deliver their Upkeeps against whichever targets the First and Second Waves had failed to destroy. Yet it was almost certainly a serious mistake, which cost lives, to schedule the Reserve Wave's take-offs so late, so that they encountered thoroughly awakened German defences, and also within a few hours the beginnings of daylight.

When Brown, who had completed just six previous operations, crossed the Dutch coast just behind his fellow Canadian Lewis Burpee and followed by Bill Townsend one minute later, it should already have been plain that all three should head for the Sorpe. Cochrane wrote later, in self-justification: 'There was nothing wrong with the priorities, but a lot was wrong with the run of luck which prevented so many aircraft reaching the Sorpe.' In truth, however, direction also broke down, and likewise in considerable degree communication.

The Reserve Wave had no leader, and its Lancasters were plying their courses independently. Cochrane's headquarters had no idea who had already been shot down. Why was there no designated leader for the Sorpe attack, to perform the same role as Gibson at the Möhne and the Eder? The bleak conclusion must be that 5 Group's planners had little expectation that Wallis's bombs would succeed against the earthen dam. If that is so – which no one ever admitted – it was cruelly irresponsible to send men at such risk to attempt such a mission. It may be suspected that the principal role of the Reserve Wave was to provide insurance for the attacks on the Möhne or the Eder – masonry targets for which Upkeep was designed – if Gibson's group failed to breach them.

The architects of Chastise had long assumed that only once in the war could Germany's dams be attacked with Wallis's weapons. After Upkeep had been demonstrated in action the

Germans would deploy nets, balloons and guns, such as would render it unthinkable again to send low-level bombers against such targets. Cochrane and his planners reasoned that they must attack what they could, as best they could, knowing that the opportunity would not recur. Ken Brown was heading east towards the Möhne, near the night-fighter airfield at Gilze-Rijen, between Breda and Tilburg, when F-Freddie ran into the same belt of flak and searchlights that had troubled other crews earlier in the night. During the First Wave's passing, some German light flak batteries could not shoot, because safety barriers around the guns made it impossible to depress their barrels low enough to engage the Lancasters passing deck-high. Three hours later, however, the barriers had been torn aside, and the 37mms fired furiously. Brown's gunners relieved their feelings by blazing back. Grant McDonald, a carpenter's son from Grand Forks, British Columbia, who was manning the rear turret, later described the impotence and frustration of meeting heavy fire when flying at low level: 'You can't go down and you can't go up … so you just had to plough through it.'

F-Freddie survived, and raced onwards. Dudley Heal, Brown's navigator, said: 'We flew over this green countryside on this fine starlit night at roughly 150 feet – it all seemed so strange, everything looked so attractive. Everything was so green over Holland and again over Germany, and everything stood out – churches, pylons, river and canals. Twice, on the way there, we saw a Lancaster go down, a mile or so to port of us. But we kept going. We were lucky.' Others were not. At 0153 Brown's crew glimpsed an explosion, and a ball of fire which briefly rose as the crippled Lancaster struggled for height. Then the red shambles dipped to the ground, and a further massive explosion marked the detonation of its Upkeep. At 0230, when Grantham tried to direct Burpee towards the Sorpe, his wireless-operator failed to respond. It appears to have been his aircraft, S-Sugar, which F-Freddie's crew had seen go down almost forty minutes earlier.

The Germans believed that this Lancaster, rather than being hit by flak, fell victim to a searchlight positioned on a tower at Gilze-Rijen airfield. Its beam caught the cockpit of S-Sugar almost head-on, dazzling Burpee as he approached at sixty feet, and apparently causing him to hit a tree, plough through a plantation and finally crash into a disused garage on the airfield. A hundred yards further, said a German eyewitness – a night-fighter wireless-operator waiting to be scrambled – and the wreck would have struck the searchlight tower. The ensuing blast, as the Lancaster's bomb exploded, almost blew the watching Luftwaffe men off their feet. The rear turret was the only part of the aircraft found intact at daybreak: the gunner, F/Sgt. Gordon Brady, still occupied his seat and was not visibly injured, but had been killed by the shock of impact.

Gibson and his men left the Eder convinced that Maudslay's Z-Zebra had crashed within a few miles of the dam, destroyed by the explosion of its bomb. The squadron commander believed that he had seen some part hanging broken below the aircraft even before it made its run. Be that as it may, Maudslay and his men were able to fly much further than their comrades had imagined before fate overtook them. It is difficult to suppose that their aircraft was badly damaged, because in that event some or all of the crew would have baled out during the fifty minutes between Z-Zebra's release of its Upkeep, and its rendezvous with nemesis.

Instead, all were still aboard when, just after 0230, German gunners near Emmerich, on the Rhine less than three miles short of the Dutch border and a hundred miles from the Eder, heard a Lancaster approaching from the south-east, and trained their weapons even before achieving visual contact. When they opened fire the plane banked hard to starboard, its rear-gunner Norman Burrows returning fire. Perhaps a dozen light flak 20mms, similar to those mounted on the Möhne, engaged the low-flying bomber almost horizontally, until a flash and a sheet of flame showed that an engine or fuel tank had been hit. Z-Zebra crashed at 0236, the crew being killed instantly. Its story between the unsuccessful Upkeep drop and the crash at Emmerich remains an enduring mystery. Henry Maudslay was twenty-one. His estate was valued at £21,536, perhaps as much wealth as all the rest of 617 Squadron's casualties left together. Apart from provision for an annuity for his old nanny, though she had pre-deceased him, the money was divided among close relations. It is possible, even probable, that like so many of his contemporaries, he found his wartime grave without having known the love of a woman.

At 0232, Bill Ottley's Reserve Wave crew, still outbound from Scampton, was briefly ordered by 5 Group to attack the Lister dam, a message which his wireless-operator acknowledged. This was Jack Guterman, scion of a Jewish family that had fled Poland. He later worked in his father's Surrey accountancy business while developing a passionate interest in music, literature and art. Jack displayed talent as a painter, and devoted every spare hour at Scampton to his latest work, writing to his sister early in May: 'My "Gethsemane" is progressing and flavours of Fra Angelico, the Italian Primitive ... I get so thrilled about it that I cannot get it out of my mind and rush back to do odd things to it. I believe it will turn out to be my *chef d'oeuvre*.'

At 0234, after a hasty rethink at Grantham, 5 Group again signalled Ottley, to divert C-Charlie to the Sorpe. This time

Guterman failed to respond. The Lancaster was coned by searchlights and hit by flak north of Hamm, where three hours earlier Gibson's section had suffered some bad moments. With the inner starboard engine on fire and flames streaming past the fuselage, Fred Tees in the rear turret believed himself a dead man, because they were far too low to jump. 'I'm sorry boys,' said Ottley over the intercom. 'We've had it.' The plane ploughed into a field north-east of Hamm, where its Upkeep exploded. Ottley and the others died almost instantly, one man grotesquely impaled on a tree stump.

By a one-in-a-thousand freak, however, Fred Tees staggered from the blazing wreck, badly burned and concussed, but alive. Sixteen-year-old Friedrich Kleiböhmer had hurried from his home a mile away to explore with a schoolmate the excitements of the crash site before it was cordoned off by the authorities. Seeing a shadowy figure approaching, Friedrich said jovially to his friend, 'This is a Tommy, for sure.' And so indeed, to their astonishment, the dazed figure proved to be. Tees raised his hands in the air and said hesitantly, 'British.' He was a twenty-year-old barber's son from Chichester in Sussex, who had swapped turrets that night with front-gunner Harry Strange – too late for the squadron's typed order-of-battle list to notice, but in time to save his own life.

Friedrich's friend hurried off to telephone the police, while the teenager, armed with a hunting rifle handed to him by a farmer 'in case he tries any funny business', escorted his prisoner towards a farm. They passed two homes still completely darkened, where they did not stop because the young German 'did not want to surprise the people at night with an Englishman'. They watched another bomber pass in the distance, between Heessen and Ahlen, and Friedrich noticed Tees gaze longingly after it.

Then they glimpsed a girl Friedrich knew sitting in a farm-house window, and he called to her: 'Elli, open up, I have an Englishman here who cannot walk any further.' Once inside they sat the prisoner down on a chair in the kitchen, where in the light they saw leaves in his hair, and shocking burns on his hands. He showed by gestures that his ribs were causing him pain. The young people's mood of wonderment was supplanted by a display of bitter hostility when Elli's father came in and exploded: 'These are the bastards who are ruining everything for us!' Tees was too shocked, in too much pain, to respond to this. He sat slumped, shivering, until after half an hour a policeman arrived in a taxi and removed him to the nearby town. The Englishman was fortunate to be humanely treated. It was not unknown for British and American aircrew who landed in Germany to be killed out of hand, especially in cities suffering at their hands. Tees spent the rest of the war first in a hospital, then in a PoW camp.

At 0228, Bill Townsend was directed to the Ennepe by a signal from Grantham, while Ken Brown and Cyril Anderson were ordered to make for the Sorpe. Brown in F-Freddie was already close to the Möhne, and thus had to fly only a few miles further to the Sorpe, where he and his crew embarked upon the same cautious exploration that Joe McCarthy had undertaken two hours earlier. They could see loose masonry atop the dam, fractured by McCarthy's Upkeep. The Germans below, in the village of Langscheid, experienced a new surge of bewilderment and terror as the huge bomber overflew them again and again, groping for an approach. For their part, the nerves of Brown and his crew cannot have been improved by having seen Burpee's aircraft go down just an hour earlier. The night was ebbing. The Germans should have been thoroughly awakened to the menace to their dams; at any time, fighters might intervene. Unless Brown and his crew hurried, the protective darkness would be gone, with probably fatal consequences for their homeward passage. Moreover, patches of mist were now drifting across the lake, which thickened on each of three runs that they made – and aborted. Only hilltops were clearly visible; valleys were hidden beneath greyness.

Brown decided to take the extraordinary risk of dropping incendiaries to provide a ground datum for his approach. Stefan Oancia, the twenty-year-old bomb-aimer, Canadian son of Romanian immigrants, pushed them through the flare chute on their next run, repeating the process on the one after. It is hard to overstate the risks involved, and thus the courage required to do this. Every moment that the fliers lingered in this most hostile environment in the world, their peril increased. Yet only at 0314 were pilot and bomb-aimer belatedly satisfied that they had the right course and line. On their seventh run along the Sorpe, Oancia released their Upkeep with extraordinary accuracy, some fifteen feet below the crown of the wall, less than twenty yards from its centre. They all heard the explosion that followed, and circled to view its effect. They saw more rubble on the parapet, close to that left by McCarthy's mine. To their bitter disappointment, however, the dam still stood. Dudley Heal, the navigator, said later, 'We had done the best we could.' It was a remarkable best. Rear-gunner Grant McDonald spoke for five others when he urged Brown down the intercom, 'Get out of here! Go on – get moving!' At 0323 they sent the signal 'GONER78C' – another report of failure. Thus ended the last attack on the Sorpe that night: Germans on the ground likewise logged the end of the British air assault on their dams at the same hour and minute.

The explosion of Brown's bomb caused even greater fear and fury among the villagers beneath than had that of McCarthy almost three hours earlier. Every window shattered in the power station below the dam, and thousands of roof tiles were blown off houses in Langscheid. Telephone lines went dead. Local party leaders arrived – 'gentlemen of the brown colour', who ordered immediate repairs. They also reported grimly to the Langscheiders that the Möhne was broken.

Meanwhile Bill Townsend and his crew in O-Orange endured an ordeal by flak. Having taken off from Scampton after midnight, over both Holland and Germany they found the defences thoroughly awake, and had what the pilot later called 'a very, very nasty' flight, during which they several times glimpsed the enemy gunners below, and Townsend 'threw that heavily-laden Lancaster around like a Tiger Moth' to escape their attentions. There were urgent debates on the intercom between crew members about the best means of evading successive clusters of flak and searchlights ahead – outbreaks of democracy such as would have been unthinkable in Gibson's aircraft. At 0222 they received a wirelessed order from 5 Group to attack the Ennepe dam, thirty miles south-west of the Möhne. Twenty minutes later they glimpsed a lake below … only no lake was marked upon their maps.

O-Orange's wireless-operator had missed picking up the NIGGER and DINGHY signals, and thus their first intimation that the Möhne had broken was the sight of civilians beneath the aircraft, scrambling onto the roofs of houses to escape a torrent rushing past them. There was excited chatter among O-Orange's crew as the Lancaster passed the Möhne itself, where they saw 'a great stream of water rushing out of the breach in the dam and rolling down the valley below the wall … a horrifying, frightening sight', in the words of Australian navigator Lance Howard, a pre-war commercial traveller. Yet wireless-operator George Chalmers said, 'I don't think we ever gave a thought to the people who would be drowned – at least I never did. There was a job to do and we just did it. We spent a little time looking at that, then carried on to our own target.'

Mist made navigation difficult as they hastened onwards, finding it ever harder to establish pinpoints in the hills and valleys below. On belatedly locating Townsend's designated target, the Ennepe dam, his crew made three aborted runs before finally releasing their Upkeep. They found the experience especially nerve-racking, because while all the Lancasters vibrated as their mines counter-rotated, in O-Orange the sensation was so powerful – probably because its load had been badly balanced – that they became fearful for the airframe. The bomb sank fifty yards short of the dam wall at 0337, inspiring in their rear-gunner an uncomplicated relief that they had seen the last of their monstrous burden. There were no eyewitnesses, and no damage was done. Thus, to this day argument persists about whether Townsend indeed attacked the Ennepe or, instead and more probably according to German sources, the nearby Bever dam. As they raced home through the growing daylight and thick patches of mist, Lance Howard later wrote, 'I remember looking back at the country ... and seeing a miasma rising from the ground, looking to me after the events of the night like all the evil in the world manifesting itself.'

Meanwhile P/O Cyril Anderson, piloting Y-York, experienced misery and frustration. Having taken off last, at 0015, just as Gibson reached the Möhne, over Holland he and his crew were driven off track by heavy flak, and thereafter suffered ongoing navigational difficulties as the early mist thickened. They received radio orders from 5 Group to make for the Sorpe, but at 0310 found themselves still far from the target, with dawn just an hour away, and their rear turret having become unserviceable. Anderson made the wholly under-standable decision to turn back with his bomb towards Scampton. Y-York thus became the only aircraft other than the damaged W-Willie and H-Harry voluntarily to return without attacking. A fellow-pilot later cruelly accused Anderson of being 'in a hurry to get to bed'. Yet the difficulties and losses that befell the Reserve Wave were overwhelmingly the fault of commanders and planners, who scheduled their take-offs far too late, placed no one in charge at the targets, and sought to impose repeated changes of objective while crews were in the air or already dead. Ken Brown, like Joe McCarthy, made herculean efforts to perform a mission impossible, and has received less recognition than the dedication of himself and his crew deserved. Cyril Anderson and Bill Townsend merited sympathy.

And so they went home – or not, as the case might be. At 0036 Les Munro, perhaps the most frustrated man of the night, landed W-Willie at Scampton, taking the risk of bringing his Upkeep home rather than jettisoning it in the North Sea, because the message had been repeatedly drummed into them that this was the RAF's most secret war weapon. Since Munro was bereft of communications, he was obliged to descend onto the runway unannounced, avoiding by only a hair's breadth collision with Geoff Rice's H-Harry, which was making a simultaneous approach a few feet above him, hastily aborted by the pilot. Rice had reached Scampton well before the New Zealander, but felt obliged to circle overhead until he had almost emptied his fuel tanks, and lowered the undercarriage with the aid of the emergency bottle. In the absence of a tail-wheel, his crew took up crash positions behind the main spar before he made his approach; they contrived to land safely at 0043, just before the Möhne was breached.

Munro described himself as 'bitterly disappointed', while Rice was crestfallen about having 'made a complete balls of things'. Charles Whitworth, who drove out to H-Harry after its crash-landing, urged the pilot to cheer up as he gave him a lift to debriefing. Sir Arthur Harris said, when he arrived at Scampton later and heard Rice's story, 'You're a very lucky young man.' The disconsolate airman hung around, awaiting news of the fate of others.

Gibson and most of his crews were fortunate enough to experience return passages less eventful than had been their outward flights. Dave Shannon ignored the plotted routes to dash for a straight run home: they had been briefed that they could push throttles through the gates, and so they did, defying the usual restrictions designed to safeguard engines from overstrain. As Les Knight and his crew passed the Möhne, they were astonished by how dramatically the reservoir level had already fallen, exposing great tracts of sand on its shores. They found themselves once again under fire, from a surviving flak gun below the dam. Later, as the light grew, the crews could see on the ground below cattle, chickens, people, such as had been invisible on their way to the dams. Richard Trevor-Roper in Gibson's rear turret had already fired almost twelve thousand rounds. He used up the last of his ammunition on a village they overflew just before leaving Germany to cross Holland. Fred Sutherland likewise blazed the front guns of N-Nuts at a train crossing in front of them as they passed a small town, and was childishly delighted by the firework effect of his coloured tracer ricocheting off the locomotive.

When L-Leather approached the Dutch coast in growing daylight, Dave Shannon said, 'I think we've got to do a quick whip out in case the coastal batteries are waiting for us.' To gain speed, he climbed to eight hundred feet before pushing down the nose and hurtling over the sand dunes, then hugging the salt waves all the way home. Gibson waxed poetic in his description of the first glimpse of the glorious emptiness beyond the coast: 'the calm and silvery sea – and freedom. It looked beautiful to us then – perhaps the most wonderful thing in the world. Its sudden appearance in the grey dawn came to us like the opening bars of the "Warsaw Concerto" – hard to grasp, but tangible and clear. We saw the yellow sand dunes slide below us silently, then we were over the sea with the rollers breaking on the beaches and the moon casting its long reflection in front of us – and there was England … We would be coming back.'

As the crew of Ken Brown's F-Freddie had feared, they found daylight creeping up on them as they crossed Holland on their way back from the Sorpe, and drifted slightly off track. They paid the price by meeting a coastal flak belt in which the pilot was momentarily blinded by searchlights, and the plane hit. When at last they broke through and were safely over the sea, navigator Dudley Heal crawled back to inspect the damage, and found the rear fuselage badly holed. His wireless-operator said, 'Hey skip, come on back and crawl in and out of the holes.' Brown passed the controls to flight-engineer Basil Feneron and clambered over the main spar to marvel that they were still airborne – and homebound.

Micky Martin was the first of the attackers to reach Scampton, immediately followed by David Maltby. Having left the Dutch coast at 0153, just after the Eder erupted, they landed at 0319, followed by Joe McCarthy six minutes later. The American's flight back had been almost as alarming as his trip to the Sorpe. T-Tommy became badly lost, and was obliged to circle for some minutes near Dülmen in search of lakes that eventually provided a pinpoint. They were caught by searchlights, which the pilot said his gunner had been unable to engage because of 'stoppages in the rear turret'. They then heard an ominous thud under the aircraft, though unaccompanied by fire or obvious damage. They flew on to reach the North Sea unscathed, passing low and apprehensive over a British convoy – naval gunners were notoriously undiscriminating in their choice of airborne targets. A painfully bumpy landing at Scampton, starboard wing down, revealed a burst tyre, and a spent cannon shell lodged in the fabric of T-Tommy beside navigator Don MacLean. The cause of the earlier thud became obvious – as did their amazing luck.

Dave Shannon, first of the Eder attackers, touched down at 0406, followed by Gibson nine minutes later, Les Knight at 0420, Cyril Anderson at 0530. Ken Brown landed safely at 0533, followed by Bill Townsend last, at 0615, after a troubled passage home in which he had to shut down an engine. As he approached Scampton he found that hot oil from the front guns had smeared the windscreen so badly that he was obliged to peer out of his side window to judge the approach. O-Orange bumped repeatedly before finally settling clumsily to earth. As its weary pilot climbed down from the cockpit, a gruff, unfamiliar voice demanded to be told how the trip had gone. Townsend irritably told the questioner to wait until debriefing. Sir Arthur Harris was unamused.

617's squadron commander had been longest in the air, over six hours. This was shorter than the duration of many Bomber Command operations, especially those to Berlin. But considering the extraordinary exertions and stresses to which Gibson had exposed himself that night, it was remarkable that he was still standing, and coherent, after completing the routines of switching off booster pumps, master-engine cocks, ignition, then clambering down from the cockpit. Geoff Rice noticed that the CO's hair was plastered with sweat as he walked into debriefing amid the hum of excited, enthusiastic voices. A 5 Group commander later recalled the mood of those moments: 'I can still smell the early-morning smell of stale cigarette smoke, taste the strong bitter tea out of heavy NAAFI cups' – on this occasion stiffened by whisky for those who wanted it – 'and feel the sinking feeling … as an ops room orderly chalked up the time landed of one after another of the names on the board, and the clock clicked round towards the ominous moments when the remaining blanks in the "time landed" column would mean that there would be no return.'

Gibson and his comrades were congratulated by Harris, Cochrane and Wallis, who had left Grantham at 0400 and arrived at Scampton in time to hear the First Wave crews being debriefed. All the pilots agreed that the use of VHF voice control over the targets had been critical to success: thereafter, throughout the balance of the campaign it would become commonplace for Main Force operations to be directed by a 'Master Bomber'. Shannon claimed to have made a nine-foot gap in the Eder, later widened by Knight, but the evidence was against him.

Harry Humphries said to Maltby, 'Hello, Dave. And how did it go?' The pilot answered, 'Marvellous. Absolutely marvellous. Water, water everywhere – wonderful, wonderful. A terrific show, Adj., absolutely terrific. I have never seen anything like it in my life.' Then he added sombrely, 'Hoppy's bought it. Shot down over the target … Some didn't even get there.' At debriefing an exultant Micky Martin recited again and again the wartime RAF's favourite expression of enthusiasm: 'Gunners did wizard work … Navigation and map-reading wizard.'

Everything had been wizard, except that he complained with theatrical indignation to Mutt Summers about the impertinence of the Möhne flak gunners who had made holes in P-Popsie. Joe McCarthy said he thought that issuing the gunners with continuous tracer had been a mistake. Many pilots observed that if the dams had been defended by dazzling searchlights, deck-level attacks would have been impossible.

A few of the exhausted young men went to bed, but many partied for hours. There was resentment that while the officers' mess was kept open, with beer and whisky freely available, the sergeants' mess was not likewise provided. Some Waafs were dragged from their beds to join a conga through Charles Whitworth's house, until the celebrants sank into unconsciousness wherever they chanced to collapse. Harry Humphries' last recollection of the night, or rather early morning, was of 'several stubble-chinned, bleary-eyed aircrew types croaking out dirty songs about the Germans around the mess piano'. Barnes Wallis lingered in the debriefing room until persuaded by Charles Whitworth to take advantage of a bed in his quarters. The engineer was devastated by 617 Squadron's losses, a price for the fulfilment of his vision such as he had naïvely never contemplated.

As morning filled the Scampton sky, they waited in vain for the others. Young left the Eder just ahead of Gibson, but had failed to land. As it became plain that his aircraft would not return, uneasy jokes were made about dinghies – how Melvin must have decided yet again to paddle home. This time, however, there was no raft, no rescue. A-Apple had been caught by flak at the cruellest moment, in the cruellest place, hit in the tail just after crossing the coast to the North Sea and safety, fifteen minutes before Gibson crossed in the same area. Young's

comrades were convinced that, on the way home as on the way
out, he made the mistake of flying too high. His Lancaster
crashed on a sandbar, leaving no survivors. His wife Priscilla
wrote to a former fellow-flier after his death: 'Melvin was apt
to lack confidence in his own abilities, and the only way in
which he could gain assurance was to tackle some hard job.'
Breaching the Möhne had been the supremely 'hard job'. It
seemed especially harsh that he had thereafter been required
to linger over Germany for a further hour after bombing, as a
stand-by leader at the Eder if Gibson went down; that thank-
less role cost him his life.

There were tears among some of the ground crew, and from
WAAF mess staff who saw so many places laid at the breakfast
tables that would not be filled. Indeed, Scampton's permanent
personnel displayed more emotion than did the aircrew, who
were exulting in the success of the mission and their own
survival. Warriors become familiar with the awful, guilty thrill
of emerging alive from experiences that have been the deaths
of others. Sid Hobday said laconically, 'We were used to it, of
course.'

The destruction of the Möhne and Eder dams had been
achieved at the cost of eight aircraft lost out of nineteen that
took off, almost half the force that got as far as Germany,
including both flight commanders. Of the losses, only one, or
possibly two – those of Hopgood and Maudslay – were directly
attributable to the hazards of using Barnes Wallis's weapon.
However, the punitive overall toll – for 617's casualties were
heavy even by the standards of Bomber Command – represented the cost of conveying Upkeeps to Germany in the
moonlit conditions that alone made Chastise feasible. Luck
was at least as important as skill in determining which aircraft
were lost on the night of 16/17 May: Hopgood, Young and

Maudslay were among the best and most experienced pilots in 617 Squadron; what happened to Young, especially, along with others shot down by flak, could as easily have befallen Gibson, Martin or others. It seems less remarkable that almost half the attackers were lost, than that so many got home.

No definitive judgement is possible about the reasons some bombs did not run true: pilot or bomb-aiming error, or – in the cases of Hopgood, Martin and Maudslay – battle damage might have played a role. Bomb-aimer Len Sumpter said sensibly that nobody on the ground or in the air 'could prove anything really, whether the bombs were on target or not, in all the excitement and the firing and the noise of the planes going over'. Wallis may well have been technically correct that a single Upkeep, in the right place, would breach a masonry dam. We know that of eight Upkeeps dropped at the Möhne and the Eder, five failed to fulfil the engineer's exacting demands upon their deliverers. However, both the complex modifications of the Lancasters, and the Wallis bombs, appear to have worked almost perfectly on every run. For this Avro and Vickers technicians, together with the ground crews at Scampton, deserve more applause than they sometimes receive, alongside Wallis himself.

Yet no weapon which needed to be launched against an enemy target with such extreme precision could be considered fit for common usage. What 617 Squadron achieved against the two dams which they breached represented a freak of history, skill, courage – and fortune – such as no responsible commander could ask a similar force to repeat. Guy Gibson wrote in his logbook: 'Led attack on Möhne and Eder Dams. Successful.'

8

The *Möhnekatastrophe*

1 'A WALL OF WATER, BLACK AS COAL'

The power of water in human affairs, for both good and evil, is almost beyond imagining. The breaching of the Möhne dam unleashed upon the valley below a hundred million tons from the reservoir, led by a tidal wave sometimes forty feet high, which swept down the Sauerland with primeval force, bearing death and devastation at a speed of six yards a second, creating floods that eventually extended a hundred miles, to the confluence of the Ruhr and the Rhine. When the Lancasters of 617 Squadron first approached the midnight darkness nearby, air-raid sirens wailed across the region, causing many families to take refuge in the cellars that they used for shelters. Local authorities and inhabitants failed to respond as the real circumstances, the real threat demanded, because it was beyond their imaginations to conceive of the mighty Möhne's collapse. They feared bombs, blast, flame; but not water in inconceivable volume.

The dam warden, *Oberförster* Wilkening, 'wasted precious minutes', in the words of the subsequent official report, attempting to get through to the local headquarters in Soest using his direct telephone link, before recognising that the line was severed. He then ran to the station of the Ruhr-Lippe railway and delivered a hopelessly belated warning at 0110, followed by a second message twenty minutes later, reporting 'an impending flood catastrophe'. Power lines to the post office that received his calls had been cut, and thus the switchboard staff sought to spread word by candlelight, with pitiful and hopeless sluggishness, as the flood rushed towards them much faster than the tidings.

A local resident first saw what to him appeared 'peculiar mushrooms of smoke', but were in reality clouds of spray, rising above the forest below the Möhne. There followed 'a swooshing and a rolling, like the sound of trains', but because this was incomprehensible, it was also unfeared. 'Suddenly tumultuous screams echoed through the night.' Householder Willy Kaufmann experienced an electrifying instant of understanding: 'That is the Möhne reservoir which is approaching.'

The breaking of the Ruhr dams created widespread disruption. Yet hours before the floodwaters reached major factories, mines, workshops, they swept through valleys where their victims were cattle, horses, pigs. And people. Destruction of the dams would win applause around the world, partly because causing the collapse of such masses of masonry seemed to carry none of the moral baggage of attacks on cities, which already troubled some thoughtful people. Yet on 17 May 1943, 617 Squadron killed by inundation far more civilians than had any earlier RAF raid in Sir Arthur Harris's 'Battle of the Ruhr', or indeed any operation since the British strategic air offensive began.

Karl-Heinz Döhle had watched with his father from the garden of their home at Himmelpforten, a tiny community set amid woodlands two miles below the dam, what they came to call 'the big birds of doom', and had heard the thunderous explosions that followed their passage. In pursuit of a better view of the extraordinary spectacle, they ascended a low hill named Koster's Coast, above the thirteenth-century Baroque church of Porta Coeli and the cloisters of a ruined Cistercian monastery beside it. From this high ground, which saved their lives, they heard a deafening roar to the east, from the direction of the dam. Within seconds, the church vanished beneath foaming water. Its tower remained visible for a further few moments, then collapsed beneath the flood, its bells delivering a final peal, abruptly stifled. The priest, Father Berkenkopf, probably drowned in the cellar where he had taken refuge from a feared rain of bombs, though some accounts suggest that it was he who tolled the bells for the last time. Himmelpforten ceased to exist: even the rubble of its ruins was borne hundreds of yards down the valley.

On and on westwards surged the waters towards the hamlet of Günne, bearing on their billow fallen timber, flotsam and wreckage of all kinds. When daybreak revealed trees stripped of bark after being repeatedly lashed by driving debris, Karl-Heinz Döhle thought they looked like white ghosts. In Günne, three hydro-electric plants and the clubhouse of the local shooting association vanished, taking with them the bodies of thirty inhabitants, some of which were later recovered miles downstream. The tracks of a narrow-gauge railway were torn asunder, twisted into fantastic shapes among the trees. In narrow places of the valley the flood rose to a height of almost forty feet, then fell somewhat as it extended deadly fingers ever more widely across flatland between the hills.

Nobert Kampfmann, a tobacconist in the town of Neheim, received a telephone call from his brother-in-law in Niederense shortly after 0100. In tones of panic he cried, 'The Möhnesee is coming, raise the alarm and get to safety!' Kampfmann rang the police station, where his warning was sceptically received. Nonsense, he was told. If anything like that was true, the station would have been informed through official channels. Other callers who raised the alarm were likewise rebuffed, so that hundreds of local people lingered in their shelters, fearing bombs, instead of fleeing before the racing deluge, adopting the only course that offered salvation – a dash for higher ground. The local police afterwards admitted that they had received warnings before the flood arrived, but said that it was impossible usefully to respond, when the entire local population was obediently underground.

Elisabeth Lingenhofer, whose husband was absent, serving with the Wehrmacht, had been lodging for a fortnight in the hamlet of Niederense with her sister-in-law and three young children, in a half-timbered house where they had just finished putting up curtains. That Sunday evening she had cycled into Neheim to see one of Germany's first-ever colour movies, *Die goldene Stadt* (City of Gold), at the Apollo cinema. The sounds of gunfire and explosions at the Möhne caused her to waken the children, and at first take refuge in their cellar. Many others in the communities below the dam did likewise – and perished. The Lingenhofers were fortunate enough to emerge from underground on hearing strange noises above. Just in time they glimpsed the rising waters, and hastened to the upper floor of the house, which was soon surrounded by 'a raging lake'. They cried in vain to neighbours for succour; then, as the level continued to rise, retreated to the attic.

There also water soon swirled, for a time attaining a height of twenty-four feet above ground level. Elisabeth frantically tore away tiles to make an opening, through which they clambered up onto the roof-tree. The children cried hysterically, clung to their mother. The grown-ups prayed. Many who appealed to the Almighty that night were disappointed, but the Lingenhofers' prayers were answered. While hundreds of houses in the valley collapsed, their sturdy old building some- how held up. Elisabeth said later that it seemed a miracle, with every joint creaking and groaning, the structure repeatedly battered by passing debris, dead cows. They saw fantastic sights, such as a small timbered building floating by, in which a candle still flickered behind a window. A mass of wooden planks and poles, washed into the current from a flooded lumberyard, surged amid the chaos.

After many hours, when at last the waters began to subside, the mother put her youngest daughter to sleep in a washtub before descending to the ground floor – with some difficulty, in the absence of the vanished staircase. All their furniture had been swept away, and a thick layer of sludge coated every surface. Elisabeth said fervently: 'Never again will I make a home in a valley below a dam.' Among those who had done so, 181 citizens of Neheim perished. Yet the heaviest human forfeit for the *Möhnekatastrophe*, as local people would forever call it, was paid by nearby people – almost all of them young women – who were not Hitler's followers, nor even his citizens, but instead his slaves.

Throughout the war, the Nazi industrial machine was hampered by a shortage of manpower, mitigated by exploita- tion of PoW labour, especially French, together with vast numbers of conscripted workers from Germany's new Eastern empire, dominated by Poles and Ukrainians. Over half of all Soviet forced labourers, and one-third of Poles, were women. Some five million foreigners, including concentration camp prisoners, eventually toiled for the Third Reich. Of these an estimated 170,000 perished, from Allied bombs or their oppressors' cruelties.

Many of the East European workers, including some who worked in the Möhne valley, were leased out as domestic servants, in fulfilment of a Nazi policy intended to sustain the morale of the *Heimat* amid the stresses of war. Most, however, were held in camps in conditions of chronic privation. When the first shipment of Ukrainians to Germany was organised in February 1942, Nazi official Graf Kajetan von Spreti reported irritably that because of the panic among the deportees about what awaited them in the Reich, he had found it necessary to hang two Jewish women and one Jewish man for allegedly spreading false rumours.

Most of the recruits were aged between sixteen and twenty, abducted from schools and institutions. Walter Letsch, a civil servant in the Reich Labour Ministry, reported: 'The mass recruitment and deployment of young Russian women and girls in the Reich is necessary for reasons that are decisive for the outcome of the war. It is indispensable.' Many women were initially employed on farms. It was there discovered that they worked impressively harder than their German counterparts, which caused large numbers to be diverted into industry. Employers welcomed the freedom to keep their workers toiling for as many hours as they could stand; there were no issues about the health and safety of so-called '*Ostarbeiterinnen*'.

The plight of those living in the barracks of the Möhne valley, just upriver from Neheim, was grim. We know their names, from Nazi personnel rosters – Anna Petrova, Valentina Filipova, Lisametta Klimova, Kathrina Tapikova, and hundreds of others – but little more about them. They were confined by night to huts in wired compounds, atrociously nourished, then dispatched at daybreak to labour for ten to twelve hours in factories and homes. One woman worker wrote: 'We were fed upon turnip soup every day, together with a kilogram of bread a week. That was all. We were permanently half-starved, subsisting on waste from German kitchens such as vegetable peelings which we cooked and ate in secret.' A minority of local people treated the workers with kindness and sympathy. Most employers, however, regarded them with less sentiment than they accorded to their beasts of burden. Fraternisation was strictly forbidden; sexual relations were punished, even when coercion could be proved. The only consolation for these tragic people was a belief that their compatriots in Ukraine, Poland and Russia were suffering even worse things. In the compounds outside Neheim, at least they seemed to be quarantined from violence – until the coming of the horrors that befell them in the early hours of 17 May 1943.

Ferdi Droge was a sixteen-year-old apprentice living in Neheim, who heard rumours of the fate of the Möhne and hastened through the darkness to check the river. As he ran he suddenly heard 'terrible cries' from the site of the nearby camp, a little way upstream, where over a thousand Eastern labourers were lodged: 'I had seen these women and young girls every day, from the windows of my own workshop, as they went to work in the town's factories.' Neheim contained works that manufactured light electrical components and torpedo fuses, also an aluminium plant. A steep hill above the valley in which lay the barracks of the foreign workers channelled the rushing water towards their compounds. Most inmates, trapped behind barbed wire, had no hope of escaping the doom which struck them with the roar of an express train.

From higher ground two hundred yards distant, Droge glimpsed a 'wall of water, black as coal, heading towards me at least forty feet high, within it fragments of huts tumbling on top of each other, entangled with screaming people. A few tiny lights' – lanterns, perhaps – 'spun amid the whirling timbers until they were snuffed out by foam.' Droge ran as he had never run before, glimpsing a flash that momentarily lit up the landscape as water short-circuited the massive United Utilities electricity transformer. He became one of a throng of desperate people hastening up a hillside thoroughfare named Peace Street. Only when they seemed to have climbed high enough to save their own lives did they pause, breath exhausted, to turn and gaze upon the horrors being endured by others less fortunate: 'the crashing, swooshing, cracking and death screams'.

Some women foreign workers swept by, clinging to the broken timbers of their barracks building until these smashed with crushing force into the wall of an office building: 'the screams died'. Frau Gustel Schulte, who had found safety up a local hill called the Wiedenberg, listened to the cries of helpless women still trapped inside an intact hut which sailed past on the torrent. At last it struck a bridge with a force that killed or drowned the occupants: 'The screams of those women still ring in my ears ... terrible – just terrible.'

A certain number of the foreigners survived, because a few Germans behaved well. Ukrainian Darja Moros felt lifelong gratitude towards an elderly, limping guard known to them only as Robert. As a mob of terrified women crowded at the narrow gate in the fence of their camp, to which he lacked the keys, Robert produced pliers, then wrestled with the stubborn wire to open paths of escape, hastening away such prisoners as could wade through, urging, 'Quickly, quickly,' while the flood rose to their chests. In those awful moments, people survived who kept their footing to struggle to higher ground, while all those who slipped and fell perished – including Robert. Moros wrote: 'On the mountain behind the river Möhne we stayed until morning, drenched, half-naked, freezing. Then we were led to the house of a farmer named Herr Plesser, who gave each of us a piece of bread and a mug of milk. Afterwards, we were sent back to work again.'

Daybreak revealed the corpses of hundreds of Russian and Polish women, strewn amid their pathetic possessions. More victims were exposed six months later, when the Möhne mill-race was drained. To this day, nobody knows the exact number of those who died: many bodies were simply designated *Unbekannter Russe?* – 'Unknown Russian?'. The toll of foreigners included some male French and Belgian PoWs, and totalled at least eight hundred, among them the respected Russian camp doctor, Dr Mihailova, together with her children. Fifty Poles and a similar number of French PoWs were buried in a mass grave in the local cemetery, described on their common tombstone as having died in *'le rupture des barrages de la Ruhr'*. Another mass grave, later decorated with Polish colours, was marked merely as the resting place of 479 unnamed Poles and Russians. Some Ukrainian women who survived were found in the morning clinging to trees, stark naked and bleeding, the clothes torn from their backs in the violent passage downstream.

A vivid reflection of the mentality of the Nazi officials conducting the later recovery effort is that their report records with dismay the damage and German loss of life, but refers to the dead slave labourers only to note that construction of the camp in which they had been housed, now destroyed, had cost a million marks.

The flood that ravaged the camp rolled on to wreak havoc in Neheim town. Willy Kaufmann told his wife to fetch a cushion and their small son, while he himself seized the emergency packs of valuables and essentials that every German family within range of Allied bombers kept to hand. They paddled through water that was already ankle-deep to escape from their house, then hastened towards the nearby hill, Kaufmann constantly urging his wife, 'Higher, higher.' From their position of relative safety the fugitives saw the first houses of their neighbourhood collapse beneath the rushing tide, the cracking noise causing Kaufmann to make an odd comparison 'with [the sound of] a dentist, prising out a tooth'. They threw restraining arms around a neighbour, frantic for his missing family, who sought to return downhill to save them.

A hundred French PoWs, who fled their hutments after merciful guards unlocked the doors, mingled with local people. Some fifty others, however, were working a night shift at the aluminium plant. These hapless men were locked into the plant's air-raid shelter when the alarm sounded, and never released. Next day when their tomb was pumped dry, prisoner André Guillon described seeing the corpses of his drowned comrades suspended in grotesque postures, 'trapped between the heating pipes and the roof'.

Among many terrible sights that night, some were merely bizarre, such as that of two pleasure boats that customarily plied the Möhnesee, and somehow survived a tumultuous rush through the dam breach. Thereafter they cruised serenely, 'almost elegantly' down the valley, amid whirlpools and debris. The nightwatchman of a local light-fittings factory, Kaisers, passed the home of its proprietor and found there only Anna, a Russian conscript worker who served as the family's cook, weeping piteously. Her husband, whose photograph stood beside her bed, had perished serving with the Red Navy, and she often sobbed both for him and for the homeland for which she yearned. Now, she told the watchman with Slavonic fatalism, 'Tonight I shall join my husband.' Locking herself in her room, she remained there to die, while the rest of the house's occupants fled the flood. Thirteen-year-old Hermann Kaiser watched from the hill in his pyjamas as 'the water torrents, wide as the Mississippi, heaved through the Möhne valley with the sound of twenty-five express trains'. His father endured with fortitude loss of the family house and devastation of the region, but nursed lasting grief about the loss of a cherished stamp collection.

A parish priest, Joseph Hellman, described how his church, which stood safely above the flood, was designated as a mortuary, which soon held some two hundred bodies: 'men, women and children, from infant to greybeard, were laid out to be identified ... Some corpses, disfigured by water and injuries, had to be registered as unidentified. It was terrible to look upon the dead as they lay there, some with clenched hands and faces distorted with fear. [Others] wore peaceful expressions, as if they slept. Some had passed into eternity clutching in their hands a cross or rosary.'

The tidal wave rolled onwards past Neheim, overturning the locomotive and trucks of a coal train near Wickede/Ruhr station, tearing tracks from their bedding, sweeping away the rail embankment. In the little town of Vosswinkel, inhabitants were first alerted by the pounding hooves of a stampeding cluster of horses which galloped up its main street ahead of the flood. In Wickede, a community of 3,300 people, Karl Brockmeier's twelve-year-old daughter chanced a few days earlier to have found a long rope in the street, which she bore home as childish booty. Now, as the family trembled for the survival of their house amid rising waters, her father threw one end of this suddenly precious lifeline to his neighbours the Rennenbaums, who tied it to their chimney. Brockmeier then slid along it 'like a tightrope artist', with his daughter in front of him, ten feet above the raging flood. His wife still lingered on in their attic, however, clutching her baby. When the entire house collapsed a few minutes later, mother and child were precipitated into the torrent. Frau Brockmeier, bleeding in a score of places from injuries inflicted by a cascade of roof tiles that fell upon her, somehow clung both to the child and to a heavy timber. Six hours later the pair were rescued, in a deliverance unusual that night.

When the flood subsided, the barefoot Brockmeier pulled on two odd boots that he found in mounds of debris, then waded into the ruined home of another neighbour, named Henke. There he found a mother and her four children lying dead in a morass of mud, furniture, wreckage. Another neighbour, Herr Bauer, stood on a nearby street corner, surveying the shambles among a disconsolate cluster of other survivors. This, he said bitterly, was what '*der Führer*' had brought upon them. A loyal Nazi listener promptly reported him to the Gestapo, who detained him for interrogation. When the Brockmeiers scoured the debris of the surrounding area for relics from their home, they found only a sewing machine, caught in a hedge beside a factory. This was soon afterwards stolen by scavengers as desperate as themselves.

Hanna-Maria Kampschulte was a sixteen-year-old living in Wickede with her mother, her eighty-two-year-old grandmother and two brothers – her father was away fighting with the Wehrmacht. They had listened fearfully to the distant explosions from the direction of the Möhne, then heard the aircraft-engine noise recede. As silence once more descended upon the night, they left their shelter in the cellar and returned to bed, assuming that the All-Clear would soon sound. Instead, the children were hastily reawakened by their mother, who heard terrifying noises. Hanna-Maria peered out of the window and screamed, 'Water, water!'

The family hurried up to the attic, and watched in horror as the flood rose to the upper storey of the house, just below their feet. The younger boys wept; their despairing mother exclaimed that they seemed fated to die together. Young Willi Kampschulte suddenly cried out: he had seen the home of their neighbours, a couple named Meier, parents of five children, suddenly vanish, together with its occupants. Hanna-Maria watched a big pear tree crash, its trunk tearing away the frontage of another neighbour's home. With the wall gone, for a time its inhabitants were exposed in plain view, on their knees upstairs, praying by candlelight: 'Then an enormous wave swept by, causing the house and its occupants to collapse on top of each other.'

A slight jerk beneath their feet indicated to the Kampschultes, alternately sobbing and praying, that their own time had come. The house was falling, the water rising, so that soon they were forced to cling onto the inside of their roof to keep their heads above the flood. 'My mother said, "Hannchen, you can swim, and Willi can try too; maybe you can save yourselves. What will Papa do without us?"' Then they found that the entire attic and roof were being swept downstream, still with its human flotsam attached. As they drifted through the darkness Hanna-Maria smashed some battens and tiles, so that she could thrust her head through and breathe freely. She pulled her whole body upwards, and for a time clung to the roof tree. Then this too began to break up. The rest of the family had vanished. The young girl found herself alone in the water, grasping a tier of wooden floor planking.

'Around me were doors, furniture, barrels and boxes, floating alongside bellowing and dead cattle. Suddenly I heard shouts for help in the mist. I called, "Who's there?" It was our neighbour Frank Lohage, who had his mother and a Fräulein

Neurath with him.' Then these others drifted away into the darkness, and Hanna-Maria found herself swept over a weir, losing her grip on the timbers to which she had been clinging. She grabbed in vain for a passing telegraph pole, then achieved successive brief moments of salvation clutching a wardrobe, a bedstead, finally a door. She drifted over a railway embankment, before becoming caught up in a roadside avenue of pollarded willows. Hoarse from shouting hopelessly for help, she clambered into the branches of one of the trees, only to suffer persecution from an unhappy, lonely cow that thrust its head above the current to lick her feet.

Desperately cold, she nonetheless stripped off her waterlogged clothing, the better to brave the torrent when, as she every moment expected, she was once more carried away. Instead, however, the first glimmerings of day showed the waters falling; the tidal wave had rushed past; was now far beyond. Around 8 a.m., a soldier waded out from the shore – one man among almost a thousand dispatched to the area a few hours earlier for rescue and recovery duties – and gathered in his arms the filthy, exhausted, bedraggled teenager. Hanna-Maria was borne to safety in a farmhouse where she was washed, put to bed, and later reclothed from the farmer's grandmother's wardrobe. Later in the day, a horse-drawn cart took her four miles back to Wickede. With her home gone, she trudged instead to that of an aunt and uncle: 'They thought that they were seeing a ghost. In my black clothes, I seemed an apparition.' Next day, the 18th, her father arrived on leave from the front to confront unspeakable family tragedy: 'I was the only thing left to him.'

The two forlorn figures rode on borrowed bicycles to Fröndenberg, where they eventually identified the corpses of Hanna-Maria's mother, grandmother, and youngest brother

Udo. Brother Willi was never found, though his father disin-
terred a succession of corpses from unmarked graves in a vain
search for the child, numbered among Wickede's 118 dead.
The local official report observed grimly: 'The rescue and identification service was handicapped by the fact that bodies were
covered in mud, badly mutilated and rapidly decomposed in
the hot weather. Immediate burial was insisted upon, to avoid
the risk of epidemics, though no such outbreaks have been
reported.'

On 30 May Helene Schulte wrote a long letter to her brother
in Berlin, describing the mood in Fröndenberg after the flood
receded:

> I still cannot get over the fact that our government is capable
> of such negligence. Experts must have recognised the danger,
> and could have prepared people accordingly. The flood
> waters hit us two hours after the Möhne was breached. What
> might have been achieved in those hours? So many people
> and cattle didn't need to die … Now we sit here in the midst
> of our rubble, recoiling from the stench of decay, having
> barely escaped with our lives. The Sorpe has developed
> cracks, too, and might one day 'bless' us as well. Thus we find
> ourselves living upon a powder keg: we awaken with a start at
> any shout, or on hearing running feet … During the first
> weeks frustration was compounded by the fact that some
> people's cellars were still flooded. [Whenever a new air-raid
> alarm sounded] many such folk bolted like startled deer,
> racing for the safety of higher ground …
>
> Those who did not witness [the events of 17 May] cannot
> imagine this roaring, raging monster. Under the moonlight,
> everything was clearly visible … [Next day] my hands
> became bloody and torn from scrabbling through the mud in

our cellar in a quest for valuables and papers ... Our bank savings books were found later, by chance, in a field. Until Friday evening [the 21st], we lacked clean drinking water ... Sanitary conditions became terrible. We all received [food] hand-outs, but what good is asparagus, if you cannot cook it? [The soldiers who came to help with relief work] are normally stationed in cities that suffer bombing. They do not show much sympathy for the plight of people like us, saying that they have seen much worse things in Essen, Dortmund etc. They admit, however, that water wreaks more havoc than fire ... I cannot describe how awful things have been. Downstairs looks worse than a thieves' kitchen. Upstairs we now have wonderful unsullied views ... because all the trees that once hid the prospect have been swept away. Fondest regards, your Helene.

At lunchtime on 17 May, a thousand passengers on the 1.30 train from Hagen to Dortmund suffered the terrifying experience of seeing the Herdecke viaduct, thirty-four miles below the Möhne, collapse before them as the locomotive slowly rounded the curve onto its approach. Several people hastily threw open doors and leapt down onto the sagging tracks. Schoolgirl Lotte Buerstatte described 'a vast tumult of water, raging, gurgling, mingled with trees, roofs and animals. Then there was a cry in the carriage: through the window we saw the brickwork of a pier crash into the flood.' Somebody pulled the emergency cord, and the engine halted just short of the breach, almost a hundred feet above the flood: 'We jumped down from the carriage then ran, ran and ran back along the embankment.' After a long delay while the crew discussed what to do, with painful caution the engine inched away from the chasm, pushing its carriages safely back onto solid earth.

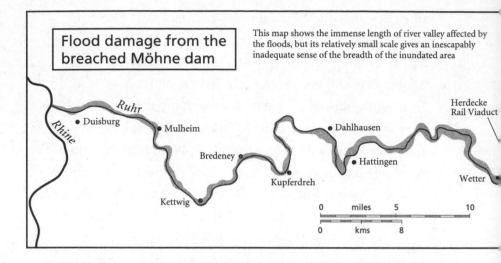

Flood damage from the breached Möhne dam

This map shows the immense length of river valley affected by the floods, but its relatively small scale gives an inescapably inadequate sense of the breadth of the inundated area

On 9 June 1943, police chiefs in the areas affected by the Möhne collapse compiled a preliminary report on its consequences. Almost six thousand cattle and 625 pigs had perished; over four thousand hectares of agricultural land had been flooded, sludge-caked and rendered uncultivable; almost a hundred factories had been more or less seriously damaged, thirty-three slightly so; forty-six road and rail bridges had been wrecked or damaged; bathetically, the report included the destruction of thirty-three clusters of beehives. The final death toll was estimated at 1,294, or possibly as high as 1,579 – several hundred people were still missing. Over a thousand houses had been destroyed or badly damaged, in addition to vastly more that were swamped, their inhabitants' possessions ruined. When, a week after Chastise on the night of 23 May, more than eight hundred heavy bombers of Harris's Main Force raided Dortmund, 650 people died, and the work of the firefighters was grievously impeded by lack of water, caused by the draining of the Möhne.

*　*　*

(Above) The only known image of a 617 Squadron Lancaster taking off from Scampton for Operation Chastise on the evening of 16 May 1943.

(Above right) The wreck of a Lancaster, believed to be that of Dinghy Young's 'A-Apple', on a Dutch sandbank on the morning of the 17th.

(Below right) One of the most celebrated, and also most cynical, images of the wartime bomber offensive: Sir Arthur Harris stands listening to the debriefing of Gibson's crew back at Scampton, as he claimed public paternity of the dams attack he had opposed beforehand and privately dismissed afterwards. Ralph Cochrane stands on his left.

Survivors after the raid: Fay Gillon sits with David Maltby on her right, Les Munro on her left; in foreground Richard Trevor-Roper, David Shannon on right.

King George VI at Scampton with Gibson on his right, Charles Whitworth left.

Further aircrew who served in the attack. (Back row left to right) Sutherland, Kellow, O'Brien. (Front row left to right) Hobday, Johnson, Knight, Grayston.

Fred Tees, the rear-gunner who miraculously survived the crash of Bill Ottley's Lancaster, and the letter sent to his mother (wrongly identifying him in the front turret).

Reference :-
DO/6/43

No. 617 Squadron, RAF Station,
Scampton, Lincs.

20th. May, 1943.

My Dear Mrs Tees,

It is with deep regret that I write to confirm my telegram advising you that your son, Sergeant F. Tees, is missing as a result of operations on the night of May 16/17th., 1943.

Your son was Front Gunner of an aircraft detailed to carry out an attack against the Mohne Dam. Contact with this aircraft was lost after it took off, and nothing further was heard from it.

It is possible that the crew were able to abandon the aircraft and land safely in enemy territory, in which case news will reach you direct from the International Red Cross Committee within the next six weeks. The captain of your son's aircraft, Pilot Officer Ottley, was an experienced and able pilot, and would, I am sure, do everything possible to ensure the safety of his crew.

Please accept my sincere sympathy during this anxious period of waiting.

I have arranged for your son's personal effects to be taken care of by the Committee of Adjustment Officer at this Station, and these will be forwarded to you through normal channels in due course.

If there is any way in which I can help you, please let me know.

Yours Very Sincerely,

Guy Gibson
Wing Commander,
Commanding, 617 Squadron, RAF.

Mrs. E. Tees,
23, St. James Rd.,
Chichester, Sussex.

W Derrick
Fred Tees

The flood, having ebbed from its first terrible surges, courses through Neheim, below the Möhne.

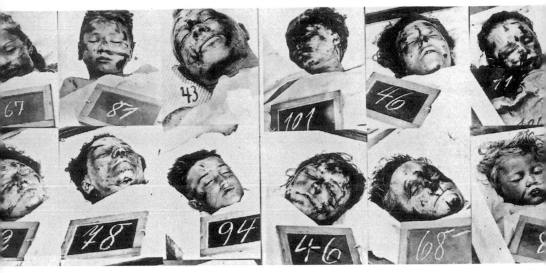

In Fröndenberg, more than 160 unidentified dead were numbered with blackboards.

Housewives struggle to salvage furniture and household goods that were swept away by the *Möhnekatastrophe*.

Hitler's armaments supremo Albert Speer (above centre) was initially appalled by the destruction of the Möhne, but organised effective repairs. The wooden scaffolding that fronted the dam all summer of 1943 emphasises how vulnerable the works would have been to even a moderately accurate conventional bombing attack.

Arthur Harris, the glowering and by then embittered former C-in-C of Bomber Command, with his wife and daughter on their way to South Africa soon after the end of the war.

Operation Chastise becomes a British national legend in the hands of director Michael Anderson, whose 1955 movie *The Dam Busters* became the most popular British war film of all time. In this supremely evocative shot, Barnes Wallis (right) stands alone, reliving a host of memories during shooting at one of Teddington's experimental ship tanks, where he had carried out some of his most important experiments twelve years earlier, during the development of the 'bouncing bomb'.

Wallis with Redgrave, who played his role on screen, and made him famous.

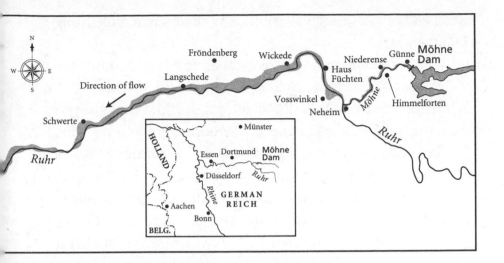

Civilian casualties below the Eder were far less severe than in the Möhne valley, because the area was more thinly inhabited: just forty-seven people are thought to have died between the Edersee and the industrial city of Kassel. The valley was shallower, which dispersed the flood and weakened its initial force. Moreover, the breach in the dam was perhaps twenty feet narrower than the 250-foot gap in the Möhne, which made the outflow slower. Three-quarters of the Edersee's estimated 202 million tons of water escaped, but this loss was spread over forty-eight hours. Alarms were telephoned and broadcast with greater speed and to more effect than at the Möhne, though some fugitives escaped with only seconds to spare.

Fourteen-year-old Heinz Solzer was at home in Affoldern when the mayor of nearby Hemfurth rang to report the Eder burst: he told the family to hasten to alert their community. The boy ran outside with a friend, Heinz Heck: the two grabbed their bikes and pedalled furiously up the road banging on windows and shouting, 'The dam's burst! Save yourselves – get up the hill!' The chain of Heinz's bike repeatedly slipped off as he made his own escape to higher ground, from which he watched houses wrenched asunder by the torrent, collapsing into the water.

There were some miraculous escapes. In Mehlen, down the valley from Affoldern, August Kötter heard the alarm. A soldier on leave from the Russian front, his first thought was for a prize pig in a stall beneath the house. He told his wife, 'We've got to see if we can get it upstairs.' Dashing below, he liberated the squealing captive, which then 'ran up the steps into the kitchen as though it had done such a thing a hundred times before'. Their children had already escaped to higher ground, but the parents knew that they themselves had left it too late to escape. They heard a deafening bang as a blast of water blew in their front door, then within seconds Herr Kötter found himself immersed up to his chest, watching their pig swept away through a window, bellowing in panic.

The couple fled first to the upper floor, thence to the attic, where they stood nursing terror, intensified when they saw a neighbour's house vanish. Then the front of their own home collapsed, leaving them clinging to a crumbling pile of masonry and timbers. Kötter lashed together a primitive raft, composed of a ladder and heavy roof timber, with a steering oar made from the pole that 'we always used to raise the flag on the Führer's birthday'. He had just contrived a safety rope for his wife from a washing line when the raft capsized and began to drift away. Kötter threw himself into the water and managed to drag his wife, rigid with terror, back through the bathroom window to a fragment of the house still standing. Later in the morning, when the water subsided, they were rescued by a boat from Hannover-Münden. Even their pig was eventually returned alive, from a farm six miles down the valley. Kötter thought: 'At the Russian front, in the fighting and the cold, I was spared – yet back here in the *Heimat*, I came close to a horrible watery grave.'

Heinz Wiesemann, in Lieschensruh, fancifully likened the approaching wall of water to a hay-cutter mowing the valley. He saw three horses drift past, tethered to a wooden hay manger. After fleeing from their home, he and his family spent the rest of the night in the waiting room of Buhlen station, among a crowd of other refugees sobbing for lost relatives. At dawn the water ebbed sufficiently for them to return to their houses, most of which had survived. Among many dreadful sights, young Heinz recoiled in disgust from the corpse of an elderly man from Affoldern lying in a field of rye that had now become mud: worms crawled in the victim's waterlogged grey beard. In the valley immediately below the dam, 103 buildings and fifteen farms were recorded as 'destroyed', together with large livestock losses. Four power stations below the dam were temporarily shut down.

The hours before the Eder floodwaters descended to Kassel provided time enough for people in low-lying areas to gather their valuables and escape to safety. Yet by afternoon there was six feet of water in this city of 220,000 people, affecting 2,500 homes. Beyond damage to public buildings, industrial and military installations, 1,130 of the city's dwellings were recorded as having suffered damage. Power supplies continued uninterrupted, but when the waters receded almost two thousand people were left homeless.

2 'CLOSE TO A SUCCESS'

The authoritative modern German Potsdam *History of the Second World War* describes the impact of Chastise upon the population of the Möhne and Ruhr river valleys as 'catastrophic'. The chief minister of Westphalia, Karl Friedrich Kolbow, wrote: 'The destruction of the Möhne dam goes beyond all imagining. In the lower Möhne valley and Ruhr valley between Neheim and [the] Hengsteysee [reservoir] there is total devastation ... no one would have believed back in 1911, when the Möhne was built, that for their homeland it would prove more of a curse than a blessing.' The best modern estimates are that between 1,300 and 1,500 people died as a result of the floods precipitated by the dams raid, of whom around half were PoWs or foreign workers, most of them the women confined above Neheim. Many of the fields in the Möhne valley, polluted by sludge, remained untilled when the war in Europe ended two years later.

A German woman once described how, in 1945, she and her daughter were raped by Russian soldiers, their home despoiled. She said: 'It seemed so wicked, when we knew that we had done nothing wrong.' That was how almost all those who endured the 1943 *Möhnekatastrophe* viewed their own plight – as a monstrous injustice thrust upon the innocent. Yet Portal, Harris, Cochrane and indeed Winston Churchill would have said that the horrors that befell those in the path of the floods were consequences of the total war Germany had unleashed, and which the Nazi leadership professed to welcome; that the bitter fruits of Chastise represented, for the people of the Sauerland, a portion of the price Germans must pay for Hitler.

The citizens of every nation engaged in the Second World War, notably including Britain, had already suffered terribly from air attack. The inhabitants of Churchill's cities had become well used to dragging corpses, many of them those of women and children, from wreckage contrived by the Luftwaffe. The people of Russia were daily suffering losses that by 1945 dwarfed those of every other belligerent.

The critical question, however, in the eyes of Chastise's British architects, concerned not the human tragedy it precipitated, but its impact upon the warmaking capacity of the Third Reich. Albert Speer, Hitler's armaments supremo, was awoken in the early hours of 17 May to be given the news of the Möhne disaster. At dawn he took off by air for the Sauerland, and overflew the floods before landing at nearby Werl. His first reactions were shock and alarm at the implications for Ruhr industry, together with astonishment at the havoc wreaked by 'a mere nineteen bombers'. He reported personally and in the bleakest terms to Hitler, who was enraged, not least by the failure of the Luftwaffe to frustrate the attack. On this occasion the Führer's anger was less irrational than usual, because it was indeed remarkable that Gibson's aircraft had been left free for almost two hours to manoeuvre around the dams, only a few minutes' flying time from night-fighter airfields, without incurring interference.

Within days, however, Speer recovered his nerve. Granted plenipotentiary powers by Hitler, he performed one of his accustomed feats of management and improvisation, to restore production and start upon repair of the dams. On 18 May he formed the *Einsatzstab Ruhrgebiet* – Ruhr Area Special Unit – which was authorised to use fifty thousand foreign Todt Organisation workers both to clear immediate wreckage and debris, and to address repairs to the dams. Work on the latter began at the Möhne on the 28th. Skilled Dutch, French and Italian artisans played a prominent part in carrying out the huge task of closing the breaches, which employed up to eighteen hundred workers, and was uninterrupted by the RAF.

Speer wrote: 'That night [of 16/17 May], employing just a few bombers, the British came close to a success which would have been greater than anything they had achieved hitherto with a commitment of thousands of bombers.' He was bemused by the RAF's diversion of effort to the irrelevant Eder, and even more so by the failure to interrupt repairs on the Möhne with conventional air attack: 'A few bombs would have produced cave-ins at the exposed building sites, and a handful of incendiaries could have set the wooden scaffolding ablaze.' On 23 September, a mere eighteen weeks after the Möhne was breached, the dam was once more sealed, in time for the reservoir to catch the autumn rains – though at a cost that has been estimated as equivalent to several billions in modern money. Closing the Eder took slightly longer.

Despite the above, it seems mistaken altogether to dismiss Chastise's impact on Hitler's war effort, as some iconoclasts have sought to do. Bruce Page wrote in a devastating 1972 *Sunday Times* article: 'The truth about the Dams raid is that it was a conjuring trick, virtually devoid of military significance.' Chastise indeed failed to inflict upon the Ruhr's industries the decisive blow that Barnes Wallis and Sir Charles Portal had sought. But the immediate havoc exceeded that achieved by any previous single air attack on the region. For weeks after the flooding, affected areas reported disruption of production caused by water shortages at shaft mines, coking plants, smelting works, power stations and armament factories. A dearth of gas was among the secondary effects. In May 1943 German coal production fell by over 800,000 tons. Almost one-third of this shortfall was directly attributed to the Möhne collapse, together with a further loss of 154,000 tons caused by damage or flooding at miners' homes, which caused wholesale temporary absenteeism. On 20 May, police reported an outbreak of panic in Fröndenberg following a wild rumour that the Sorpe had broken, and indeed terror of some new torrent persisted for many months across the Sauerland.

The ports of Duisburg and Ruhrort were closed for several days. Sea-mines, stockpiled at the Neheim factory where they were manufactured, had to be scrapped. Below the Möhne, every road and rail bridge for thirty miles was brought down, and it proved the work of months to restore them. Silting of agricultural land and damage to plant extended for forty-five miles. Waterworks at Westhofen, together with two major power stations thirty miles below the dam, were completely flooded. Water supplies were reduced by 70 per cent, and were not fully restored for six weeks.

The Arnsberg *Regierungspresident*'s report, composed on 24 June, described widespread damage and chaos in his area, but added: 'dark as the picture of destruction may seem, it is a pleasure to note with what vigour peasants and farmers address themselves to the task of clearing their land ... It is planned to rebuild totally destroyed farmhouses in a different style, and move them entirely out of reach of flood water, while the town centre of Neheim can be completely replanned.'

A significant cost was imposed by the mobilisation of workers, some diverted from fortifying Hitler's Atlantic Wall, to repair the Möhne and Eder and conduct other reconstruction work. There was a further diversion of resources to secure every dam in Germany against another low-level attack. Quad-mounted light flak guns, balloons, smoke-dischargers and three hundred searchlights were deployed, together with nets and mines in profusion. In May 1944, additional flak guns were rushed to the Möhne, in response to fears that the British would mark the anniversary of Chastise by reprising it.

The dams raid notably enhanced German respect for British warmaking capabilities. Helmuth von Moltke of the Abwehr, an impassioned foe of the Nazis with a British mother, wrote to his wife: 'The wicked little angels' – a sophisticated little joke about Pope Gregory I's depiction of the flaxen British slaves in a Roman market as resembling 'angels, not Angles' – 'last night thought of a new piece of nastiness, they drop torpedoes in reservoirs and cause dams to collapse ... I wonder if they have more of that kind of thing up their sleeves.'

Study of an intact Upkeep, which survived the crash of 617 Squadron's E-Easy, won the Germans' reluctant admiration. The chief of staff of the Luftwaffe's Air Fleet One spoke of 'the shock inflicted by the new specialised weapons'. Arnsberg's district president claimed to have warned in 1941 of the potential threat of an attack on the dams, which he was told by the Luftwaffe was impossible – as indeed it had been, before the creation of Upkeep. Another official report concluded: 'The attack has demonstrated that we can be hit very hard from the air by a very small force ... more serious than the impact from many [RAF Main Force] terror attacks in the Ruhr area.' Millions of Germans were obliged to confront the vulnerability of the Reich, the emptiness of repeated promises about their security made to them by Hitler, Goering, Goebbels.

While the strategic air offensive failed to break the spirit of Hitler's people, as Lord Cherwell, Sir Arthur Harris and other airmen hoped that it would, such a night's work as that of 617 Squadron imposed a trauma which it seems foolish to understate. A further judgement will be attempted below, about whether such a stroke, which imposed a heavy toll upon those civilians in the path of the floods, was justifiable in the wider context of Allied warmaking then, or seems so now, before the bar of history.

Yet on 17 May 1943, the Möhne and Eder valleys lay 'on the other side of the hill' from wartime Britain, in the midst of Hitler's Germany. In the context of those times, it was too much to expect that the young airmen who had breached the dams, at the cost of the lives of more than a third of their number, should think much, if at all, about the stricken families of the Sauerland. Before the men of 617 Squadron set forth from Scampton, they were promised by their commanders that success would contribute importantly to hastening Allied victory. They can scarcely be blamed for having believed this, nor be grudged the season of glory that now opened upon them, for having triumphantly fulfilled most of their objectives.

9

Heroes

1 GARNERING THE LAURELS

Sordid realities about thieving hands demanded that one of the first tasks for Bomber Command's ground staff on the morning following every operation was to lock the billets of those posted missing, before a 'Committee of Adjustments' officer arrived to secure their possessions. At 1000 on the morning of 18 May, Chiefy Powell led the CoA's man to one after another of the Scampton quarters of 617's fifty-six lost fliers, their progress accompanied by trucks of which every inhabitant of the station knew the significance.

Pathetic personal effects returned to the father of twenty-three-year-old P/O Vernon Byers in Pontrilas, Saskatchewan, were typical of those of most of the aircrew lost on Chastise. Apart from clothes, they included two New Testaments, three cigarette cases, a crested Ronson lighter, an unserviceable wristwatch, one book, a bottle-opener and eighteen handkerchiefs. It was little enough for a parent to receive in exchange for what was gone – all those childhood years of love, nurture, soaring hopes, such as every family cherishes.

Chiefy Powell also directed the dispatch of fifty-six telegrams to next-of-kin, to be followed by typed letters signed by

Guy Gibson, many with some personal sentence added. The CO told Fay Gillon that he was specially grieved by the loss of such pilots as Hoppy Hopgood, whom he himself had persuaded to come to 617. Bobbie-Gwen Parfitt, Charlie Williams's fiancée in Nottingham, was sent no official notification of the loss of his aircraft. Unrecorded as next-of-kin, she received confirmation of his death only on 18 August 1943, more than three months after E-Easy crashed.

Many of those who had perished proved to have bequeathed all that they owned to their parents, in the absence of wives or lovers. Melvin Young, the pilot who almost certainly broke the Möhne, made a will a week before the operation naming his father Henry as sole executor, perhaps because he lived in Britain, while Priscilla, Young's wife of nine months' standing, was in the US. The airman left £4,434 – not a fortune, but far greater wealth than most young fliers possessed dead or alive: more than ten years' pay for a squadron-leader. His widow received less than half of this money because, as Young himself recorded, she was independently wealthy. His father inherited the balance of his estate. Priscilla never remarried, and spent the rest of her long life in Boston, mostly working for musical charities. She lived to be ninety-one, her brief marriage an enigma not untypical of the wartime era.

On the morning of 17 May, after service photographers had snapped group images of the 'winning team', 617's CO quizzed the three pilots who had failed to attack on the previous night. The flak damage to Les Munro's aircraft was self-evident, and thus likewise the absence of any cause for blame. The New Zealander nonetheless felt uncomfortable making one with the squadron's party-goers through the days that followed: he felt that his own experience gave him nothing to celebrate. He had shared its many pains and travails through the preceding weeks and hours, then, albeit through no fault of anyone save the enemy, been denied its moment of fulfilment at the dams.

After hearing Geoff Rice's account of the loss of his Upkeep over the North Sea, Gibson commiserated: 'Bad luck. I almost did the same. It could have happened to anybody.' Rice participated with distinction in several later squadron operations, and survived the war after being shot down over Belgium and taken prisoner in December 1943.

The CO was less impressed with Cyril Anderson's account of his Early Return after becoming lost in Y-York: pilot and crew were summarily posted back to 49 Squadron, from whence they had come. They flew a further fifteen operations with Main Force before failing to return from a 23 September attack on Mannheim. There is no record of their sentiments about the humiliation they endured on and after 17 May.

At lunchtime that day, the BBC announced the achievement of 617 Squadron in breaching the Möhne and the Eder, quoting the Air Ministry's communiqué: 'In the early hours of this morning, a force of Lancasters attacked with mines the dams ... [which] control two-thirds of the water storage capacity of the Ruhr basin. Reconnaissance later established that the Möhne dam had been breached over a length of a hundred yards ... The Eder dam ... was also attacked and was reported as breached. Photographs show the river below in full flood. The attacks were pressed home from a very low level with great determination and coolness in the face of fierce resistance.' The sensational bulletins provided Eve Gibson with the first inkling of the mission to which her husband had been committed for the past two months. She was thus baffled when he telephoned on the afternoon of the 17th and made no mention of Chastise, instead asking about her own doings in London. This was perhaps because all phone calls from Scampton were still being monitored by security personnel. Even later, however, when they were together, he refused to discuss the dams, except to murmur about the loss of close comrades, 'It seems so unfair, somehow.' Here, perhaps, was a manifestation of that phenomenon commonplace among warriors: survivors' guilt.

Sir Charles Portal messaged congratulations to Harris from Washington, saying, 'In particular please tell the special Lancaster unit of my intense admiration for their brilliant operation.' After 5 Group's AOC had dispatched a tribute to 617 Squadron, Ralph Cochrane wrote to Barnes Wallis: 'Before reaching the end of this somewhat long but exciting day, I felt I must write to tell you how much I admire the perseverance which brought you the astounding success which was achieved last night. Without your determination to ensure that a method which you knew to be technically sound was given a fair trial, we should not have been able to deliver the blow which struck Germany.' Gibson wrote to Upkeep's begetter: 'I'm afraid I'm not much of a letter-writer, but I would like to say just this. The weapon that you gave us to deliver worked like a dream and you have earned the thanks of the civilized world.' The closeness of the two men was emphasised by the fact that the airman began to address the engineer as 'Wally' – like 'Hoppy' or 'Gibby' – while Wallis referred paternally to Gibson as 'the boy'. Though the leading lights of the RAF had long identified the Ruhr dams as desirable objectives, Wallis alone showed how the Air Staff's vision might be fulfilled. He infused with his faith the decision-makers, among whom Portal was the critical force, probably most significantly influenced by Syd Bufton and Norman Bottomley.

On the 17th also, secretary for air Sir Archibald Sinclair, Portal's political chief, spoke at a Norwegian Independence Day celebration in London's Albert Hall, describing Bomber Command as 'the javelin in our armoury'. Ironically, he awarded laurels for Chastise to 'that resourceful and determined commander-in-chief Air Chief Marshal Harris', saying nothing of that ferocious warmaker's hostility to the dams operation. Indeed, almost every published account on both sides of the Atlantic hailed Bomber Command's C-in-C as Chastise's begetter. At the Albert Hall, Sinclair went on to pay more appropriate tribute to 'those superbly daring and skilful crews who smote the Germans so heavily last night'.

Chastise was a supreme piece of theatre, and in May 1943 theatre was precious to the British war effort. On Tuesday the 18th the story made headlines in almost every newspaper across Britain. The story in *The Times* was headed 'RUHR DAMS BREACHED – DARING LOW-LEVEL ATTACK BY LANCASTERS – WALLS BLASTED OUT WITH 1,500lb MINES'. The paper wrote of 'vast damage by floods', and quoted an enemy high command statement: 'weak British air forces last night penetrated German territory and dropped a small number of high-explosive bombs on some places. Two dams were damaged, and heavy casualties were caused among the civilian population by the resulting flood.' The *Daily Mirror* headlined 'HUNS GET A FLOOD BLITZ', and the *Daily Sketch* 'FLOODS SWEEPING RUHR FROM SMASHED DAMS'. Almost every paper published the dramatic air-reconnaissance photographs taken by a Spitfire XI of 542 Squadron, showing the waters spreading for tens of miles below the dams.

Public reaction was overwhelmingly enthusiastic. The King's assistant private secretary Alan Lascelles wrote in his diary: 'The RAF attack ... appears to be one of the most cleverly-planned and best-executed enterprises of the war. If its results are anything like as substantial as they say, it may have very important effects.' A host of humbler folk joined the applause, though some were impressively sensitive to the humanitarian tragedy created by the flooding. Edward Stebbing, a medically discharged ex-soldier working in a Home Counties hospital, wrote on 18 May in his diary for Mass Observation: 'Comments (all made by men) on the RAF raid: "Wasn't it a fiendish trick, though!" "Why haven't they done it before?" "Do you think Germany will pack up now?" "I can't see it"; "Considering the size of Germany, it's just a pinprick. Because they fill a newspaper with it, it doesn't make it any more than that." There was also some discussion as to whether there are any similar dams in this country. I am in two minds about it. It was certainly a brilliant military exploit, skilfully carried out, and may do considerable damage to Germany, thus helping to shorten the war. But I can't help thinking of the people who will be flooded out of their homes or drowned.'

Twenty-six-year-old RAF corporal Peter Baxter, a Cambridge graduate, wrote likewise: 'I don't like it. The action will, I know, cause immense dislocation in this vital centre of German war industry. It will bring everything to a standstill for miles around, it will wreck communications and cause untold damage. But it will also drown thousands of non-combatants – women, children, old people and many of the wretched slaves working in Germany who have come from the occupied countries. It will render many, many families homeless and, I feel, it is calculated to create bitter rage and hatred against we British which will not easily be forgotten. In its way I consider it to be as bad as using poison gas or germs ... I know this is Total War, but are we to abandon all standards of mercy and humanity? An act like this makes us all barbarians.

I'm sorry to say that I haven't yet found anybody to agree with me. The other fellows say "The more Jerries we wipe out the better."'

In the circumstances of sorely-tried Britain in 1943, the destruction of the dams was of more significant value than were many other operations executed by the British Army, Royal Navy and Royal Air Force, more than a few of which cost the lives of innocents as well as those of enemy combatants. It is noteworthy, for instance, that a larger number of French and Dutch civilians were killed by 1944–45 Allied bombing of German V-weapon launching sites in their countries, than of Churchill's people by those same flying bombs and rockets falling upon Britain.

On Wednesday, 19 May the dams raid story still led most newspapers, with *The Times* reporting 'growing devastation in the Ruhr ... Water vital to Ruhr factories ... Kassel is now flooded and water is rising at Duisburg.' Writing from the House of Lords, however, in the paper's correspondence column, former naval officer turned Labour peer Lord Winster deplored claims made in the previous day's accounts that the raid had been inspired by 'information given by a Jewish refugee'.

The source was a former Reuters correspondent in Berlin named Guy Bettany, who sought celebrity by claiming credit for identifying the Ruhr dams as targets – with the help of a Jew. He wrote in the *New York Times* and other papers: 'It was with knowledge of their vital importance that a famous German-Jewish medical specialist, exiled from Berlin and now practising in London, asked me why the RAF had not bombed them. I was so impressed by what he told me that I wrote to the Air Ministry ... Later I had a letter thanking me for my suggestion and assuring me that it would be carefully considered. This was several months ago.' Bettany said that on hearing news of 617 Squadron's exploit he promptly phoned his Jewish friend, who expressed delight at this 'wonderful news'. Lord Winster observed in his *Times* letter: 'The result will presumably be savage reprisals against the Jews in Europe. Such statements [also] make us appear extremely stupid. Is it to be supposed that we have only heard about these dams after nearly four years of war?'

Bettany's cheap bid for fame indeed had consequences for the Nazis' foremost victims. Josef Goebbels, Hitler's propaganda chief, seized upon the Allied press claim that a Jew was ultimately responsible for the destruction of the Möhne and Eder, and broadcast it widely, in hopes of feeding the German people's loathing. In Dresden the great diarist Victor Klemperer reviewed the Nazi version of events and wrote on 21 May: 'The last few days have been dominated by the river dam business. First: the English have "criminally" bombed two dams, location not stated, very many civilian casualties. Then: it has been proved by an English newspaper article that this criminal plan was hatched by a Jew; it therefore belongs on the list of Jewish "misdeeds". The same thing was printed in the newspapers, and since then the rope around Jewish necks has been drawn ever tighter.'

On the 21st also, *The Times* published a long letter from Lord Trenchard, foremost enthusiast for strategic air attack, lauding the 'key role' of bombers in the war effort. He deplored characterisation of Bomber Command's attacks on Germany as 'raids', saying that this word implied that they were mere hit-and-run pinpricks; instead, in truth, each one was a major battle. 'The bomber,' wrote the old warhorse messianically, 'is the spearhead of the vital offensive that dominates war.' He urged that the air campaign should be doubled or trebled in weight, concluding that 'nothing must stand in the way of this'.

Some early published reports about the dams attacks suggested that the breaches had been achieved by mines dropped into the reservoirs, which were then floated onto the dams by the current. The Air Ministry strove to preserve the secrets of Upkeep, despite the wrecks of eight of 617's 'Provisioning' Lancasters being scattered across Germany and Holland for Luftwaffe inspection, together with three aircrew PoWs in its hands for interrogation. The name of Barnes Wallis was unmentioned by Allied media. His daughter Mary, at boarding school near Salisbury, nonetheless got the message when her housemistress said cheerfully after reading her newspaper, 'So your father has done it again.' Sixteen-year-old Mary dispatched a telegram to Effingham, intelligible only to the two of them: 'WONDERFUL DADDY.'

Such whispers and rumours among family and friends prompted a swift reaction from officialdom. On 18 May, Molly Wallis warned a friend in a letter:

> I write in haste to say that B's name is not to be connected with these dams. I can't really think why, unless it's the case that the Nazis find out & Effingham gets a bomb or somebody tries to scrap B. The Air Ministry rang me up late last night ... & told me that quite seriously, tho' I can hardly credit it. And they rang B up today & said that the Press had his name, & they were simply furious. Maybe it's the A.M. want to take all the credit now it's worked. Never mind, don't mention him ...
>
> Poor B didn't get home till 5 to 12 last night, only 3 hours sleep Saturday, didn't take his clothes off Sunday & was awake till 2.30 this morning telling me all about it. And then, poor dear darling Barnes, he woke at 6 feeling absolutely awful [because] he'd killed so many people.

Molly added a note four days later: 'He has gone away again [on 21 May overnight train to Scotland]. I see the point of the secrecy: there is more to do.' The floods in north-west Germany were still receding when Wallis departed to spend three days attempting to salvage the fortunes of Highball, and of the Royal Navy's Operation Servant.

Chastise took place in the midst of an exceptionally tense, sometimes ugly, combined chiefs of staff summit in the US, at which British resistance to an early D-Day on the continent exasperated Gen. George Marshall and his colleagues. On 13 May, CIGS Sir Alan Brooke had told the American chiefs that 'no major operations would be possible until 1945 or 1946, since it must be remembered that in previous wars there had always been some 80 French divisions on our side ... The British manpower position was weak.'

In the days that followed, as the outraged Americans fought back, the British gave ground. Brooke's team grudgingly acknowledged a 'firm belief' that conditions for an invasion of France would exist in 1944. Churchill felt able to cable to Clement Attlee, his deputy prime minister, that the two allies had reached agreement that Britain should have 'a free hand' in the Mediterranean until November 1943, with the objective of knocking Italy out of the war, following which a concentration and redeployment of forces would begin, for a 1944 landing in north-west Europe. Brooke wrote in his diary that he did not think the prime minister and President Roosevelt 'realised how near we were to a failure to reach agreement'.

It was in the midst of these fierce, uncomfortable discussions that Portal was able to deliver news of his service's triumph, of which the combined chiefs formally 'took note' on 17 May. Next day, Admiral William Leahy, chairman of the US chiefs, offered their collective congratulations. Considering the magnitude of the issues being discussed by the combined chiefs, however, it would be naïve to suggest that Chastise loomed large. The raid is unmentioned in Allied warlords' private diaries. In Williamsburg, Virginia, on the evening of 17 May Alan Brooke wrote wearily at the end of a day's wrestling with the Americans, amidst which Portal's announcement of Chastise represented merely a brief parenthesis: 'Another very disappointing day.'

Portal's official biography makes no mention of the dams. In Algiers, British minister Harold Macmillan, another voluminous private chronicler, merely noticed 'an immense series of telegrams – all "Immediate" or "Most Immediate" from London and Washington'. Meanwhile, none of the British aesthete diarists – for instance Harold Nicolson, 'Chips' Channon, Evelyn Waugh, James Lees-Milne – saw fit to remark upon the dams sensation. A possible explanation is that such sophisticated observers had grown wary of swallowing large propaganda claims, such as were repeatedly made for the RAF's attacks on Germany.

The achievement of 617 Squadron was nonetheless a rare event, in making headlines for British arms across America and the world. The marriage of technological brilliance with human courage and airmanship awed and thrilled tens of millions of people. It went far to explain why, in the eyes of Americans, the prestige of the Royal Air Force rose to heights that were never matched by the British Army, nor even by the Royal Navy. The *New York Times* eulogised: 'The RAF has secured another triumph. With unexampled daring, skill and ingenuity it has blasted two of Germany's important water dams ... and thereby delivered the most devastating single blow dealt from the air ... All Americans will join ... in congratulating Wing-Commander G.P. Gibson on his feat and mourn with him the loss of the eight aircraft and their gallant crews.'

When the prime minister addressed a joint session of Congress on 19 May, in the course of a rhetorical grand tour of the war, he said: 'The condition to which the great centres of German war industry, and particularly the Ruhr, are being reduced is one of unparalleled devastation. You have just read of the destruction of the great dams ... That was a gallant operation ... [which] will play a very far-reaching part in reducing the German munitions output.' The American King Features Syndicate declared: 'No one accomplishment in the war appears to have generated public enthusiasm to the extent of the mining of the Möhne and Eder dams by the RAF.'

The applause of British politicians and public was likewise unstinted, and the RAF's standard-bearers exploited it to promote support for the wider bomber offensive. When Sir Archibald Sinclair spoke at a 'Wings for Victory' rally in Edinburgh on 21 May, he echoed Sir Arthur Harris's theme song, urging that there was 'only one place in which the European war can be ended – and that is Berlin. There is only one hope of getting to Berlin without the slaughters [which] the land battles of the last war entailed, and that is the paralysis of German war power by Bomber Command.' He described the dams operation as 'an epic feat of arms', and the Möhne and Eder as 'the stiffest targets that Bomber Command had ever attacked'.

A few days later, a deluge of decorations fell upon 617 Squadron, headed by a Victoria Cross for Gibson. Harris himself telephoned his protégé at the home of Eve's parents in Penarth to reveal the award. There were DSOs for each surviving commissioned pilot who had bombed a dam – only the VC could be awarded posthumously; lesser 'gongs' were awarded to every member of Gibson's own crew. Ken Brown and Bill Townsend received Conspicuous Gallantry Medals. Ralph Cochrane remembered to pen a letter thanking the Corby and District Water Company for use of their Uppingham reservoir to train 'the Dam Busters'. This was a phrase unused by the AOC, but which in those days became a national commonplace, following the inspiration of an unidentified journalist.

At Scampton, there was some feeling that the efforts of the ground staff – eight of whom received formal commendations, though none were awarded decorations – should have been more fully recognised in the awards list. When the King and Queen visited the station on 27 May to meet Gibson and his men, there was also resentment that the visitors were entertained in the officers' mess, while the sergeants' mess was not similarly honoured, though only a minority of the aircrew who carried out Chastise were commissioned. This, however, was how things were done in the Royal Air Force of 1943.

The King was invited to review alternative crests and mottoes for the newest squadron to join the order of battle of his air force. In an echo of alleged words of Mme de Pompadour, lover of Louis XV, he opted for '*Après moi le déluge*', beneath an image of three lightning bolts above a breached dam. Thirty-four squadron medals were presented by Queen Elizabeth at a 22 June investiture at Buckingham Palace. Barnes Wallis was awarded a CBE at a separate ceremony, during which Fred Winterbotham received the same honour, though for his services to MI6 rather than his association with the engineer.

Those who received awards hastened to one of the shops that sold those slivers of coloured tape most men valued more than they admitted, and which the aircrew of 617 Squadron deserved more richly than many of those awarded in wars. At Gibson's behest Scampton Waaf Anne Fowler, who was soon to marry Dave Shannon, sewed the purple VC ribbon on his tunic. Even as she stitched in the CO's quarters, he laid bare to her the unhappiness that clung to him, in the midst of the round of celebrations in Lincolnshire and London that followed the dams triumph. His words tailed off into piercing memories of his beloved dog. Again and again he said, 'My Niggy's dead ...'

As for the squadron's dead fliers, Barnes Wallis said: 'For me the success was almost completely blotted out by the sense of loss of those wonderful young lives.' The fervour of his distress about the casualties, which persisted until his death, afflicted less deeply most of those 'at the sharp end'. Wartime bomber stations were callous places, their inhabitants inured to losses. As one of Harris's pilots observed: 'If you live on the brink of death yourself, it is as if those who have gone have merely caught an earlier train to the same destination.' Because 617 Squadron had existed for such a short time, there were few close friendships within its ranks. The spirits of the fliers revived after sleep and the grant of seven days' leave. They were young, and thrilled by the blaze of publicity and indeed adulation that descended upon them. 'We were treated like bloody gods,' said wireless-operator George Chalmers in wonder. 'It was daft. I couldn't take it in.' Gibson's navigator Terry Taerum wrote wonderingly to his mother in Alberta after being applauded as a hero when he visited the Avro factory, where Lancasters were built: 'Can you imagine me, making a speech? We were just about mobbed for autographs afterward.'

Barnes Wallis wrote to Roy Chadwick, the Lancaster's designer, responding to a telegram of congratulation: 'I think the whole thing is one of the most amazing examples of teamwork and co-operation in the whole history of the war.' This indeed it was. He added that the Lancaster was 'the only aircraft in the world capable of doing the job', which was also true. In peacetime, armed forces are obliged to do their parts served mostly by men – and in the mid-twentieth century the story was almost all that of men – of moderate talents. In the Second World War, however, belligerents suddenly secured access to the most brilliant minds in their societies, if they were imaginative enough to make good use of them, as did Britain and the United States.

The engineering support for Chastise was terrific. Compare and contrast the Luftwaffe's failure to fit aircraft fuel 'drop' tanks, to extend the range of their fighters, which could have been a game-changing gambit in the Battle of Britain. The Germans possessed the technology, but their commanders and engineers did not work sufficiently closely together to exploit it. The nearest Goering's pilots came to emulating the dams raid was a 1945 air attack on Russian electricity generators, dubbed Operation Eisenhammer, using Mistel 'piggyback' bombs and floating mines. However, in the absence of Avro Lancasters, a Wallis or a Gibson, Eisenhammer proved a failure. Chastise represented a vivid manifestation of the manner in which the Allied powers exploited the talents of their peoples much better than did the Axis.

2 SQUANDERING THE SACRIFICE

The C-in-C of Bomber Command secured exactly what he had wanted for himself from Chastise – personal credit for its achievement, which increased his authority, celebrity and – as events would show – unsackability. The *Spectator* on 21 May 1943 described the raid as 'the most destructive blow ever delivered by aircraft in a single operation. It was a perfect illustration of Air Chief Marshal Harris's policy of long-distance planning, and was then the result of deliberate previous study, exact briefing and skilful, daring execution.' Press reports of the attack, obviously based on briefings from RAF public relations officers, described how the C-in-C had played tennis on Sunday afternoon, then quit a dinner for Robert Lovett, the visiting US assistant secretary of war for air, to drive to Grantham and Scampton to watch 'his' operation unfold.

Yet while Harris welcomed the applause that Bomber Command received, he remained unconvinced of the merits of such precision targeting, as against those of incinerating cities. He wrote to Bottomley, the assistant chief of air staff: 'For years we have been told that the destruction of the Möhne dam alone would be a vital blow to Germany. Both the Möhne and Eder were destroyed and I have seen nothing ... to show that the effort was worthwhile, except as a spectacular.' This, rather than his extravagant parade of enthusiasm in 5 Group's operations room during the early hours of 17 May, reflected his considered judgement.

The acclaim conferred upon 617 impressed Harris's other aircrew less than did the squadron's frightful casualty rate. On 26 May, AVM Robert Oxland at Bomber Command set about replacing the losses by dispatching a message to all Groups, inviting them to submit names of volunteers to join Gibson's team, 'whose operational duties are not expected to be arduous'. Few were deluded. On 2 June Ralph Cochrane wrote to Harris: 'I have been trying for the past fortnight to make good the losses in 617 squadron, but so far without success. I have now heard from the other Groups that they are unable to find any suitable crews and in this Group I have only been able to select two.' Only sluggishly were Gibson's depleted ranks filled.

Harris agreed, however, to keep 'the dambusters' in being. He wrote to Cochrane: 'The value of No. 617 squadron as a specialised unit has been fully demonstrated by the destruction of the Moehne and Eder dams and ... it is my intention to keep this Squadron for the performance of similar tasks in future ... The actual operational effort ... will be small, since it will, of course, not be used on ordinary night attacks, and is therefore most unlikely to operate as frequently as once a month. I feel confident that the importance and interest attached to work of this type will make a greater appeal to many experienced aircrew than any of the alternative forms of employment open to those who have completed their second tour.'

The C-in-C fought a bitter, largely unsuccessful battle to maintain targeting choices for 617 within his own headquarters at High Wycombe, scrawling on a June memorandum following a conference held in King Charles Street to discuss the possibility of using Upkeeps for an attack on German canals: 'the proper venue for such meetings is here, not at A[ir] M[inistry]'. He nonetheless indulged the squadron's continued existence, because it conferred prestige on his entire Command; he recognised Portal's persisting enthusiasm for precision bombing; and occasional 'spectaculars' provided a means of paying lip service to the common objectives of the Anglo-American 'combined bomber offensive', using minimal resources – perhaps 2 per cent of his 1944 'heavies' – while Main Force continued its assault on Germany's cities, Harris's almost exclusive preoccupation until May 1945.

He told crews in a circular dispatched to all stations after the dams raid: 'the battle of the Ruhr is now in its third month and will go on until it has been won. It will be won when the Ruhr ceases to produce munitions of war ... If no bomb was to fall more than a mile from the aiming point it would be possible to estimate the number of further attacks required to put out of action all the chief production centres. Unfortunately a number of bombs are still falling 2, 3 and 5 miles from the aiming point, and this is delaying the victory.'

The casualties inflicted on civilians by Chastise were soon surpassed. A fortnight after the operation took place, when Bomber Command attacked Wuppertal on the night of 29/30 May, 3,500 people died. Much heavier tolls were exacted in the ensuing two years, as the strength of the RAF's and USAAF's bomber forces grew exponentially. In August 1943, when the British and Americans mounted Operation Gomorrah, attacks on Hamburg sustained over three days and nights, forty-one thousand people perished. An October raid on Kassel, which had suffered temporary flooding after the bursting of the Eder, generated a firestorm. This killed six thousand of its citizens, and 'de-housed' 123,800, more than half the city's population.

Yet it was a phenomenon of the strategic bombing campaign that even as its intensity increased and civilian casualties mounted, the Nazis deployed ever greater energy and ingenuity to mitigate its impact upon their war machine. The authoritative German Potsdam history of the war asserts: 'the experience built up ... in the industrial area of Rhineland-Westphalia in the summer of 1943 showed the way for dealing with the results of air raids in the rest of Germany. The lack of further large-scale bombing gave arms production in the west of Germany time to put right the severe damage and losses, and to form itself afresh by the time 1944 began.' In other words, even as the Ruhr's industries teetered on the brink of crisis in the wake of the March–July onslaught, which cost Bomber Command over a thousand lost aircraft including the eight which failed to return to Scampton on 17 May, Harris made the critical mistake of shifting the weight of his forces elsewhere.

The Ruhr was by far the most accessible industrial region of Germany for bombers flying from British bases. It generated coal and steel, essential elements for the Reich's entire production. Adam Tooze, a shrewd analyst of the German war economy, writes: 'Bomber Command had stopped Speer's "armaments miracle" in its tracks. The Ruhr was the choke point and in 1943 it was within the RAF's grip. The failure to maintain that hold and to tighten it was a tragic operational error. The ongoing disaster that Speer and his cohorts clearly expected in the summer of 1943 was put off for another year.'

Not only was the Ruhr reprieved, but the herculean efforts of 617 Squadron were squandered, because having breached the dams, Bomber Command chose to forget them. The post-war US Strategic Bombing Survey asserted: 'The breaking of the ... dams, though the cost was small, had limited effect.' The British official historians of the bomber offensive, Webster and Frankland, wrote: 'The effects of this brilliant achievement upon the German war machine were not, in themselves, of fundamental importance, nor even seriously damaging ... The raid did no great damage to German armament production.' Chastise caused substantial temporary disruption, but only a very limited loss of production, most significant in the coal mines. Once the reservoirs had emptied, the tidal wave swept past, the floodwaters quickly drained away.

British failure to follow through caused Speer's most perceptive biographer, Joachim Fest, to write of 'the half-hearted nature of the undertaking' by the RAF against the dams, and to cite the armaments minister's own ironic expression of gratitude for 'the powerful ally he had in the enemy's general staff'. This remark was justified: by deploying only a single squadron, Sir Charles Portal made a marginal commitment to attacking Germany's dams. His cautious or 'half-hearted' decision deserves sympathy for reasons given above. Moreover, it is implausible that much larger numbers of aircrew could have been found to deliver Wallis's bombs with comparable devotion and skill. It is nonetheless unsurprising that Portal thus secured only a marginal result. Fest continues: 'the next few days [after Chastise] revealed how little Harris had thought the programme through ... he never carried out the expected incendiary raids in the Ruhr which would have wreaked havoc because the fire brigades had no water ... The same lack of method in Allied [air] strategy manifested itself time and again.'

Why did Harris fail to order an attack on the huge reconstruction programme undertaken by Speer, to repair the Möhne and Eder before the autumn rains? In 1978, following a conversation on this issue with Barnes Wallis, I put the same question to Bomber Command's former C-in-C. He responded tersely: 'Any operation of war deserving of the VC is, by its nature, unfit to be repeated.' It was certainly true that new German dam defences precluded any repeat of Chastise. But conventional bombing could have had a devastating impact, such as Speer anticipated and feared, upon the vast latticeworks of wooden scaffolding that stood before both dams throughout the summer of 1943. That the RAF and USAAF chose to ignore these seemed to the Germans an almost divine deliverance. In consequence, in the autumn of 1943 the Möhne and Eder reservoirs resumed their role as critical water providers for their respective regions, and for their hydro-electricity generation plants.

Sir Arthur Harris must be cast as the wicked fairy of Chastise. He opposed the operation from the outset, and was disgusted that his chief, Sir Charles Portal, insisted upon its execution. Following its relative success, he took care to ensure that he himself received a credit that was undeserved, while Portal's name was unmentioned. Thereafter, his failure to mount operations to impede repair of the dams seems to reflect a petulance and stubbornness that were characteristic of his tenure of command. If Bomber Command had wrecked the repair works in a fashion that, as Speer noted, was well within its capabilities, the consequences for Ruhr industries could have been far-reaching.

Portal, the man who decreed the dams raid, played a curiously passive role here, failing to intervene as he had intervened to mandate Chastise in February 1943, now to insist upon an attack on the dams' repairs. Although there is no evidence to explain the chief of air staff's silence and apparent inertia, it may be relevant to notice the intelligence analysis that then lay upon Air Ministry desks. Bletchley Park's code-breakers swiftly accessed the special cypher used by Speer, his subordinates and local police forces to communicate about the dam disaster between 17 and 25 May. According to the British official historians of intelligence, 'the decrypts showed that the emergency was quickly controlled'.

A British assessment of 3 June 1943 stated: 'The general impression is that the police and other bodies got down to recovery and control tasks with speed and efficiency.' Knowledge that one of the major police detachments dispatched to the flooded region was withdrawn within a week, and that the special signals channel allocated to the emergency was also closed down, suggested to MI14 that the local situation in the Sauerland was back under control. The Ministry of Economic Warfare reported on 3 July 1943 that 'the [Möhne] damage did not touch the main industrial centres', though adding that the destruction of property was serious and damage, particularly to communications, 'will involve a large amount of repair work'. As for the Eder, the report stated that its most significant consequence was the temporary inundation of power stations immediately below the dam, 'otherwise the flooding affected mainly agricultural land'. MEW analysts nonetheless drew attention to the useful moral impact of 'panics and rumours' following the attacks.

Professor Sir Michael Howard dismissed Chastise in his official history of British Grand Strategy as 'a spectacular feat of skill and courage, but one whose effect on the German war effort was, unfortunately, slight'. It seems likely that Portal, like Harris, welcomed Chastise as a public relations coup for the Royal Air Force, but by July 1943 privately acknowledged that its consequences for the Germans had proved much less serious and more transitory than had been hoped; that the Ruhr dams were thus thereafter best left alone. If the chief of air staff did reach such a conclusion, he was wrong. Having once breached the dams, it was a folly not to exploit the achievement. So long as the Sorpe held, the British lacked capability to inflict absolute disaster upon the water resources of the Ruhr. Nonetheless, had the Möhne reservoir been kept empty, the outcome of Chastise could have been transformed from a temporary inconvenience into a significant blow to the Nazi industrial machine.

Just as Sir Arthur Harris showed no further interest in the dams after 617 Squadron delivered its Upkeeps, so he convinced himself that urban areas and their industries, once devastated by Bomber Command, were irreparably wrecked. In an important, wildly hubristic memorandum to the prime minister on 3 November 1943, he asserted that his squadrons had won the Battle of the Ruhr. Harris listed as 'virtually destroyed' most of the region's cities, including Kassel, Krefeld, Hamburg, Cologne, Essen, Dortmund, Düsseldorf, Mannheim, together with many others 'seriously damaged ... We claim only what can be seen in the photographs. What actually occurs is much more than can be seen. From the above you will see that the Ruhr is largely "out" and that much progress has been made towards the elimination of the remaining essentials of German war power.' Of what remained of Hitler's industrial cities, he identified Berlin as the foremost strategic target, mentioning also three smaller Ruhr cities – Solingen, Witten, Leverkusen. As for other surviving fragments of German industry, he wrote, his aircraft would 'tidy up all round when occasion serves'.

Neither Harris nor his staff ever grasped the workings of the German war economy. The interpretation of aerial photographs of damaged industrial areas was especially flawed: again and again, High Wycombe assessed targets as destroyed, when in reality buildings had merely lost their roofs. Once debris had been cleared from lathes and production lines, the energy and ingenuity of Speer's managers sufficed to revive output astonishingly quickly.

In the autumn of 1943 Harris diverted most of his Command's efforts from the Ruhr to open a new front – the RAF's so-called Battle of Berlin. In his November memorandum to the prime minister he made a series of assertions that blast his reputation as an arbiter of air strategy – the moral issues belong elsewhere: 'We can wreck Berlin from end to end if the USAAF will come in on it. It will cost us between 400–500 aircraft. It will cost Germany the war ... I feel certain that Germany must collapse before this programme which is more than half completed already, has proceeded much further. We have not got far to go.'

It is wrong to hold Bomber Command's 1942–45 C-in-C responsible for Britain's decision to embark upon 'area bombing', which was made before he took up his post. Thereafter, it was properly the duty of the chiefs of staff, and above all the prime minister, to control and, if necessary, remove Harris, should they become dissatisfied with his exercise of command. Freeman Dyson wrote of his detested chief at High Wycombe: 'His personality was not the root of the evil at Bomber Command. In himself he was not worse than Gen. [William Tecumseh] Sherman, who did evil in a just cause. He was only carrying out, with greater enthusiasm than the situation demanded, the policy laid down by his government.' Harris should have been sacked, probably in the winter of 1943, when his intemperate conduct and obsession with destroying Germany's cities were manifest; and certainly in the winter of 1944, when his defiance of Portal's orders, and of formally agreed Anglo-American targeting policies, became explicit.

As it was, however, in the winter of 1943 Harris was allowed to make an important change of objectives which the prime minister, chiefs of staff and Air Ministry made no attempt to challenge. Bomber Command launched the Battle of Berlin, an assault upon the largest and most heavily defended industrial urban area in Europe, together with other cities in its vicinity, amid chronically poor winter weather, at extreme range from British air bases. Losses were appalling, averaging over 5 per cent – one aircraft in twenty on every operation. The worst night of the entire wartime campaign fell on 30 March 1944, when 795 aircraft attacked Nuremberg, and ninety-five were lost. Among those who perished was Dick Trevor-Roper, Gibson's rear-gunner at the Möhne.

Ralph Cochrane acknowledged the failure of the assault on Hitler's capital: 'Berlin won. It was just too tough a nut.' Harris, in a letter to the Air Ministry on 7 April 1944, came as close as ever in his life to admitting that he was in deep trouble: 'The strength of the German defences would in time reach a point at which night-bombing attacks by existing methods and types of heavy bombers would involve percentage casualty rates which could not in the long term be sustained ... Tactical innovations ... are now practically exhausted.'

In the spring of 1944 Harris was saved from explicitly confronting his force's defeat in the Battle of Berlin only by the transfer of his Command – against his own wishes, as usual passionately expressed – to pre-D-Day bombing of the French rail system, under Eisenhower's orders as Allied supreme commander. Moreover, it remains a mystery why the C-in-C should have at any time deluded himself that flattening Berlin would 'cost Germany the war', even had his campaign been successful on its own terms. Portal, Harris's superior, must accept responsibility for failing to control or sack him, and for his own role in the campaign of area bombing. Portal is often said to have been the cleverest of Britain's wartime chiefs of staff. The record shows that he was also, in important respects, a weak one.

Yet, just as the air chiefs' expectations and ambitions went unfulfilled, so some latterday critics of the strategic offensive overstate their case, by suggesting that bombing contributed nothing to victory in 1945. Truth about most things is to be sought in a middle ground, rather than absolutes. It is absurd to suppose that Germany could have suffered such devastation without significant cost to its war effort. While it is true that the graphs of production of aircraft, tanks, guns and all forms of war materiel continued to head upwards even in the face of intensive 1943–44 air bombardment, in its absence they would have risen even more steeply. It is astonishing that Albert Speer was able to continue to provide arms and ammunition to Hitler's armies until May 1945, in the face of the bombing. He

was certainly not, however, capable of negating the consequences of the destruction inflicted upon cities, industries and
communications. No sustainable 'Speer miracle' was
achievable.

Industrial output was at least as severely disrupted by
air-raid alerts and their consequence – the mass flight of
workers to shelters – as by the material damage inflicted by
bombs. Eight hundred thousand Germans were employed on
anti-aircraft defence, albeit many of them teenagers or old
men, a significant number of them protecting dams. They were
allocated one-third of all gun production, almost a quarter of
heavy-calibre ammunition output, and half of all the Reich's
electronic equipment. Here was a significant achievement of
the bomber offensive – to force upon Germany a massive
diversion of resources from the ground fronts. It required the
expenditure of sixteen thousand rounds of German anti-
aircraft gun ammunition to achieve even the claimed
destruction of a British or American aircraft. Many of the
squadrons of the shrinking Luftwaffe were first recalled from
the Eastern Front to defend the *Heimat*, then shot out of the
sky over Germany by US escort fighters in the spring and
summer of 1944.

During the ensuing winter, against depleted defences
Bomber Command and the USAAF inflicted a dreadful toll
upon German cities, symbolised by the destruction of Dresden
and Leipzig. This came far too late in the war to influence the
timing of its conclusion, and merely inflicted a nightmare on
Hitler's people, to negligible strategic purpose. The great
Australian war correspondent Alan Moorehead observed that,
as he travelled through Germany and talked to its people in the
summer of 1945, he encountered 'an immense sense not of
guilt, but of defeat'. This had not existed in 1918, when the

fabric of Germany was almost untouched, and was over-whelmingly the consequence of the destruction of Germany's cities. It is a matter of personal taste, whether it was appropriate to employ the Allied air forces essentially to punish the German people for Hitler, as was done in the final months of the war.

Sir Arthur Harris long survived most of his aircrew, remaining to the end of his life in 1984 defiantly impenitent about both the policy of area bombing and his own role in it. Most of his surviving aircrew retained a rueful affection and esteem for 'Butch' Harris, as they knew him, though during the war years scarcely any of them set eyes on him. The most plausible explanation is that their own immense sacrifices would have seemed set at naught, had they renounced their wartime leader. Harris's unyielding frankness about the nature of what he did merits some respect, especially when, after the war, the principal promoters of the strategic air offensive sought prudent shelter, including Portal and Churchill himself.

Britain's former leader, in extreme old age, commented about Harris's campaign: 'I wanted it to be a rapier but it became a bludgeon … it did go on too long, and I pointed it out to the Chiefs of Staff. Such destruction tends to acquire its own momentum and those who hold weapons are apt to use them. That worried me about nuclear. Of course Harris was under-recognised at the end and so were his gallant men who suffered the heaviest casualties of all. I did something for him when I got back to power, but it wasn't much.' The old statesman, by the end, had become profoundly equivocal about the campaign for which he himself bore ultimate responsibility.

10

Landings

1 'GOODNIGHT, EVERYONE'

Following Chastise, development of Highball continued until 1949, but Barnes Wallis's bombs were denied a trial against enemy shipping – fortunately for the aircrew who might have been required to drop them. Upkeep was never used again. One hundred and twenty examples were manufactured, fifty-eight being explosive-filled, the remainder dummies. In 1945, thirty-nine of the former remained in store, and were dumped at sea. The view of both airmen and sailors was that the gods of war had smiled on Chastise as a unique operation, in which a wild and brilliant improvisation achieved tactical success against the odds. In the twenty-four hours following the raid there was some loose talk at Scampton, in which Gibson participated, about a possible immediate return to the Ruhr, to 'finish off' the Sorpe. This was mere braggadocio. It was obviously too much to ask, both of the fliers and of fortune, that further attacks should be undertaken at extreme low level against targets now conspicuous in the minds of the enemy. Even that robust young Australian Dave Shannon categorised Chastise's losses as 'unsustainable'. Italy's dams were discussed as possible objectives, because still ill-defended, but were never attacked. By the summer of 1943 it would have seemed monstrous to unleash a catastrophe upon Italian peasantry, whose government in September embraced the Allied cause.

The Germans quickly grasped the Upkeep technology, following their capture of an intact bomb on 17 May, and experience of two devastating demonstrations of its power. They nonetheless made no serious attempt to replicate its achievement. Though Winston Churchill ordered Britain's dams to be defended against such a possibility, from 1943 onwards the Luftwaffe lacked means to execute such an operation, even if it had the will. All this rendered ironic, and indeed redundant, the RAF's morbid commitment to protect the secrets of Upkeep, which persisted for more than a decade. When the feature film of 617 Squadron's exploit was shot, fantasy bombs resembling giant Dutch cheeses were fitted to the Lancasters that appeared on the screen.

Professor David Edgerton of Imperial College, London, a historian of science and technology, has written of the absurdity of the popular belief about Chastise, reflected in the movie, that a blind and obstructive wartime officialdom impeded development of bouncing bombs. Contrarily, he says: 'Wallis received massive support, to produce a weapon which was ... used on only one raid, one whose results were not as great as expected.' This does not detract from the brilliance of Wallis's conception; it merely emphasises that it proved a technological cul-de-sac.

For four months after 17 May 1943, the survivors of 617 lingered in bored and frustrated limbo, mocked by their Scampton aircrew neighbours as 'the one-"op" squadron'. They conducted three almost frivolous 'trips' to Italian targets without Gibson, dropping conventional bombs from standard

Lancaster IIIs, and on one occasion leaflets. George Holden, who in August succeeded as CO, was neither liked nor respected, though he had an impossible act to follow. On 14 September the squadron was committed to its first significant operation since May, a low-level precision attack on the Dortmund–Ems Canal with 12,000-lb bombs, from which they were recalled on meeting bad weather. The aircraft of David Maltby, a Chastise veteran, exploded over the North Sea on the return trip under mysterious circumstances, apparently victim of an accident. There were no survivors. On the following night, when the canal operation was belatedly executed, it proved a disastrous failure. Of eight aircraft committed, five were lost, including that of Holden, together with those of Les Knight, Bill Divall and Harold Wilson, survivors of the original 617, though only Knight flew to the dams.

The last-named died gallantly, holding his mortally damaged aircraft in the sky for just long enough to enable the crew to escape. Australian wireless-operator Bob Kellow described the final moments in the cockpit: 'I stood by [Les] as he firmly held the wheel and tried to keep "Nan" on a steady course, making it easier for each man to jump out … Using signs I asked if he was OK. He nodded his answer and a wry smile puckered his mouth … I gave him the thumbs-up sign … then bent forward with my head down and tumbled out into the dark Dutch night.' All seven crewmen who took to their parachutes survived, five of them reaching England having evaded capture thanks to the Dutch Resistance. Knight himself was killed when his aircraft exploded on crashing into the ground.

Of Gibson's Chastise crew, all save Dick Trevor-Roper and John Pulford – who perished in a 1944 flying accident – died with Holden on the Dortmund–Ems attack. Terry Taerum's younger brother Lorne, aged just eighteen, was also killed in

February 1945, on his first operation as a rear-gunner with Bomber Command.

Following Holden's death, Micky Martin briefly commanded 617. It quickly became apparent that, for all his courage and airmanship, an aversion to administration and paperwork rendered him an unsuitable CO. He was succeeded by Leonard Cheshire, who ranks with Gibson among the great British bomber leaders of the war. Under Cheshire's command the squadron performed extraordinary new feats of precision bombing, continued under his successors.

In 1944–45 twelve-thousand-pound Tallboy and ten-ton Grand Slam deep-penetration bombs, conceived by Barnes Wallis back in 1940, were brought into being thanks to the engineer's dramatically enhanced reputation, which yielded access to large resources. Wallis's big bombs were used chiefly, though not exclusively, by 617 Squadron. Between January 1944 and May 1945, 854 Tallboys – mostly manufactured in America – and forty-one Grand Slams were dropped. Portal, together with the Air Ministry's Directorate of Bomber Operations and Cochrane of 5 Group, sustained an enthusiasm for precision operations, which 'the dambusters' constituted the most effective instrument to carry out. Their successes in blocking the Saumur Tunnel, breaking the Bielefeld and Arnsberg viaducts, sinking the *Tirpitz* – on 12 November 1944 – as well as destroying V-weapon sites and communications targets in France before and after D-Day, constituted a remarkable record. The big bombs, and the results that they achieved, nonetheless came too late in the campaign to be considered game-changers: destruction of the great battleship, which would have been of priceless strategic value in 1941–42 or even 1943, meant little when Germany had long since lost the war at sea.

* * *

After the war Barnes Wallis served as head of research and development for Vickers-Armstrong, in which capacity he worked on many futuristic concepts, notably including 'swing-wing' aircraft. In 1950 he applied to the Royal Commission on Awards to Inventors for financial recognition of his wartime creation of 'means for executing a novel method of attack, in which a missile is released from an aircraft flying towards the target while spinning about a horizontal axis at right angles to the line of attack'. The accompanying secret report in support of the claim stated, among much else: 'A number of experiments were, in fact, made with the assistance of members of his family in his own garden.' Wallis added a personal note: 'I collaborated with Wing-Commander Gibson in evolving a scheme for training his squadron in the peculiar technique required by an attack with this weapon, and I was actually responsible for briefing him and his crews immediately before the raid.'

On 3 March 1951 the Royal Commission on Awards to inventors dispatched a cheque to Wallis for £10,000 'in full and final settlement of your claim in respect of "Upkeep" and "Highball" ... after giving careful consideration to the highly exceptional circumstances of this case'. The engineer promptly donated the money to Christ's Hospital, to create a fund to assist the education of children of RAF personnel killed in action. He received a belated knighthood in 1968, and died in 1979, aged ninety-two. His old adversary Marie Stopes sustained her animosity to exclude her son Harry from any share in her considerable fortune when she died in 1958, because of his marriage ten years earlier to Mary Wallis, though she relented sufficiently to leave a house to their son Jonathan.

* * *

Less than a quarter of those men who attacked the dams survived to see VE-Day: of seventy-seven who returned safely to Scampton on that fateful morning two years earlier, just thirty-two remained alive in May 1945. When Chastise wireless-operator Antony Stone was killed on the Dortmund–Ems operation, his mother appeared distraught at 617 Squadron's adjutant's office, demanding insistently, 'Did he suffer? Did he suffer?' Among survivors, gunner Fred Sutherland of Les Knight's crew, who had baled out over Holland, sailed home to Canada to meet an unenthusiastic reception: at Edmonton station a military policeman reprimanded him for having buttons undone. On being reunited with a childhood sweetheart, Peace River bank manager's daughter Margaret Baker, he immediately proposed marriage. The wedding took place next day, and they stayed together until her death in 2017. Sutherland, who had always loved wild places, became an inspector for the Canadian Forestry Service, dying in January 2019.

Joe McCarthy fulfilled an extraordinary flying career, completing a further thirty-four operations over thirteen months with 617 Squadron and his Chastise crew, save that in April 1944 the pilot insisted that bomb-aimer George Johnson should quit when his first child was born. 'Johnny' Johnson is now, aged ninety-seven, the dambusters' last aircrew survivor. His pilot served with the post-war RCAF until 1968. McCarthy married and raised two children, and lived in retirement in Virginia until his death in 1998.

M-Mother's bomb-aimer, John Fraser, returned from PoW camp in 1945 to be reunited with his wife Doris, with whom he had shared just one night before flying to the dams. They sailed to his Canadian homeland and raised their children, all of whom were given names commemorating 617 Squadron, including Guy, John Hopgood and Shere, the Surrey village from which the pilot came. Fraser fulfilled a lifelong ambition to qualify as a pilot, only to be killed in a 1962 flying accident.

Some others made long, happy and successful lives – notably including Micky Martin, Les Munro and Dave Shannon – and bore to the grave the laurels of their fame as 'dambusters'. Martin served as best man at his former comrade Bill Townsend's 1947 wedding, and overcame his wartime 'wild man' reputation to end his career as the RAF's commander-in-chief in Germany.

The greatest of them all failed to survive the war, of course. Because Guy Gibson became a celebrity in the wake of the dams raid, it is necessary to treat cautiously adulatory judgements that were made thereafter, by people meeting this national hero for the first time. It is nonetheless striking that Clementine Churchill, a perceptive woman, was enchanted when she encountered Gibson in Sheffield on 30 May 1943, less than two weeks after the attack: 'He is a perfect poppet – so young & modest. He is 25 but looks 18. He made an excellent speech.' Gibson told his audience: 'I had the displeasure of watching from the air your city burn during the air raids of 1940. I knew there would come a time when we should do the same to the enemy. It is said that the British can take it. We can also give it, and are engaged in blasting the middle out of Germany.' Clementine Churchill wrote: 'I thought what a nice husband he would make for Mary but Alas! He has been married for 3 years.'

Gibson's wife Eve accompanied him on a visit to Chequers, where he sufficiently impressed his host to be invited in August to join the prime minister's entourage for the Quebec Anglo-

American summit, sailing on the *Queen Mary*. Churchill began jovially to address the airman as 'Dam-Buster', just as in benign moods he called Harris 'Bomber'. The prime minister liked to show off his prominent warriors, and the Burma Chindits' unhinged leader Orde Wingate was also a member of the Quebec party. Captain Richard Pim, the naval officer who ran the prime minister's map room, described how he himself spent much of the second night of their Atlantic passage listening to Gibson's anecdotes – and judgements on the bomber offensive: 'In spite of the fact that no one realized better than he the very great damage being done to German cities by the heavy concentrated bombing, he was not prepared to subscribe to the theory held by so many at this stage of the war that bombing alone would make the Germans sue for peace. He felt that the end of the war, which was probably still two years away, would not come about until our Armies had invaded Germany through France and the Low Countries ... For one who had every conceivable decoration for gallantry [he] was most natural and unassuming.' Here is impressive evidence of Gibson's insight, contrasted with wild assurances about the war-winning capabilities of air power that his superiors were daily thrusting upon Britain's prime minister and America's president.

In the weeks that followed, the airman met most of the warlords of the Grand Alliance, and seems to have impressed them favourably: he was always better at managing relationships with big people than those with little ones. Thereafter, he embarked upon a propaganda tour of Canada and the US, during which he continued to express equivocal sentiments about the strategic air offensive. He told a Montreal audience that he was not among those who believed that bombing alone could win the war, adding, 'Of course, when I am at home I have to subscribe to that creed. But even if we cannot bomb

Germany out of the war, we can soften up the country.' On another occasion he said, 'It would be foolish to expect our bombing, devastating though it is, to result in a German collapse.' Sir Arthur Harris was outraged by remarks absolutely contrary to his own convictions, and seems never to have forgiven his young protégé. He later declined to authorise Leonard Cheshire to undertake a similar North American tour, 'after our experience with the way that they spoilt young Gibson'. The former CO of 617 Squadron never received further promotion from wing-commander, though a step to group-captain would have seemed richly justified.

On the airman's return from the US in the winter of 1943, nobody knew what to do with him. Bombing was his life, yet for months it was deemed unthinkable to permit him again to risk his life over Germany. Within the RAF, it was thought that fame had gone to his head. Anne Fowler, the Scampton Waaf who married David Shannon, had her reservations about Gibson, based on considerable intimacy with him. She observed long afterwards, a trifle ungenerously, that Mickey Rooney might have been a more appropriate choice than Richard Todd to capture his character in the 1955 movie. Always cocky, Gibson became unequivocally arrogant. His sexual activism seemed almost manic, not least with the wives of fellow officers. He renewed his intermittent relationship with former WAAF nurse Margaret North, now married to another man and with a baby son to whom he stood godfather.

He briefly embraced rash notions of a political career, professing a creed focused upon a belief that the old had got the world into war; when it was over the hopes and aspirations of the young must be recognised. He wrote of those responsible for the war: 'rotten governments, the Yes-men and appeasers

who had been in power too long. It was the fault of everyone
for voting for them.' He demanded about France and Britain:
'why had two great nations come so low?' 'If, by any chance,
we had a hope of winning this war … then in order to protect
our children let the young men who had done the fighting
have a say in the affairs of State.' In a mad moment, encouraged
by Churchill himself, he accepted an invitation to become
Conservative parliamentary candidate for Macclesfield,
though three months later he saw sense and stepped down.

He was featured as the castaway for an early edition of BBC
radio's *Desert Island Discs*, on which he told Roy Plomley
frankly: 'I'm not a highbrow by any means. In fact I can't claim
to know an awful lot about music at all. Somehow, I never
seem to have had time to do anything about it, except to listen
occasionally to something that I liked the sound of.' His first
choice of record was the Warsaw Concerto, which he said his
old squadron's officers used to play incessantly on their gram-
ophone until – here was a stab of the melancholy characteristic
of the man – 'there were very few left in the mess who remembered those that had listened to it first in the days gone by'.

And he wrote a book, which proved the second most important act of his life, after breaching the dams. The Air Ministry
posted him to a wholly incongruous role in its Air Accidents
Investigation Branch, almost certainly to provide him with a
token role while he composed, with the blessing of its large
and imaginative public relations staff, an account of his experiences as a bomber pilot. In embarking on this task, he will
have been conscious of his contemporary Richard Hillary's
memoir *The Last Enemy*, published in 1942, which we know
that he had read.

Debate persists about how much of *Enemy Coast Ahead* was
Gibson's own work. When he produced his first draft, substan-

tial sections were blue-pencilled by officialdom, on grounds either of security or the unsuitability of sentiments expressed both about air force personalities and – for instance – Jews. After Gibson's death, his widow also seems to have made some deletions, and the subsequent publishers performed editorial work. But Richard Morris is confident that the book speaks in the author's own voice – it was initially dictated – and reflects his real feelings and recollections, despite its factual inaccuracies.

Gibson appears to have received advice at an early stage from Roald Dahl, who was attached to the British Delegation in Washington when 617 Squadron's CO made his American tour. The former fighter pilot, later global best-selling author of children's books, wrote to his mother from New York on 12 October 1943: 'Dear Mama ... I've had Guy Gibson of Möhne Dam fame staying with me for the last few days and we've had quite a time. I threw a bridge cocktail party for him at which there were 2 Air Marshals & I don't know how many generals and all the pretty girls I could think of in town, just so that he could take his pick. He is the hell of a good type.' Dahl discussed with Gibson a project to make a Hollywood film about Chastise: he got as far as submitting a script to the great director Howard Hawks, though for reasons unknown nothing came of it.

Some of the shortcomings and misstatements which tarnish *Enemy Coast Ahead* reflect the fact that it was written – and intended for publication – while the war was still going on. The demands of propaganda were paramount, especially in the minds of the Air Ministry. Some strange Americanisms crept into the text, for instance references to bombers as 'ships', a word that few RAF fliers used. The author's casual condescension in writing about his own crew on the dams raid surely reflects the fact that, as human beings, they meant little to him – and were anyway by then dead.

The core of the book is nonetheless Gibson's achievement, and constitutes a significant literary legacy, given that it was the work of a young man who knew little of the world outside a bomber squadron. It became a major preoccupation of the last months of his life, the subject of one of his last letters, in which he dispatched a gift of a gold RAF brooch to the typist of his manuscript, with a note: 'I'm glad to say the book is now getting passed for censorship.'

In July he spent a day with Margaret North, now Figgins, following which he wrote to her: 'The day was perfect. I love you now and for ever.' It was their last contact. Eve Gibson's uneasy life encountered new turbulence on 16 September 1944, when she and friends were arrested for drinking outside hours at the Cardiff Squash Rackets Club. Because a national hero's name was involved, the prime minister's office was informed, either by Mrs Gibson or by police, before the case was heard. Officials warned against interfering with the course of justice, and by the time of the hearing the airman's story had anyway ended.

Roy Plomley signed off Gibson on *Desert Island Discs* by saying, 'Good luck, happy landings, and thank you.' Gibson responded, 'Goodnight, everyone.' But there were to be no happy landings. Having somehow contrived to fly on two or three raids during the summer, he was killed on the night of 19 September 1944, after acting as master bomber – airborne controller, the role he pioneered on Chastise – for a less than successful 5 Group attack on the neighbouring German cities of Rheydt and Mönchengladbach.

The return to operations of Bomber Command's most famous pilot appears never to have been authorised either by

Cochrane or Harris. Instead, this man ill at ease with humanity everywhere save flying bombers, seems to have written his own orders, which of course he should never have been permitted to do. The hapless fellow victim of Gibson's return to war was S/Ldr. Jim Warwick, 54 Base's navigation officer, who accompanied him in the twin-engined Mosquito on which the pilot, then serving as base operations officer at Coningsby, was inexperienced. Half an hour after leaving the target at 2200, their plane crashed in Holland, a victim of either flak or pilot error.

Gibson had brushed aside advice to fly a dogleg course home, above ten thousand feet, which within ten minutes would have taken the Mosquito over Allied lines – the raid took place just as Market Garden, the Arnhem operation, reached its climax. Instead he set a direct course for England, stayed low, paid the price. His body was identified only by a laundry mark. He was publicly declared missing in the British press only on 29 November. In the latter days of a world war in which so many others had died, his passing, aged twenty-six, attracted relatively little stir. Eve Gibson wrote in response to a letter of condolence from his headmaster: 'I do feel that Guy went the way he wished and that he fulfilled his mission in life.' He left no will, an estate of £2,295, and a corrected typescript.

In a 1 August letter to his London literary agent he had proposed a series of alternative titles, including *Flak-Fun-Fear*; *The Boys Die Young*; *By Fire and Water*; *Reich Wreckers* … and *Enemy Coast Ahead*. In 1945, when the book was belatedly being prepared for publication, his widow Eve asked Winston Churchill to write a foreword. The old statesman, then out of office, declined. If he assented to one such request, he said, he would be besieged by scores of other suppliants, equally deserving. Instead Sir Arthur Harris, for many reasons an inappropriate choice, provided an introduction. Mrs Gibson spent some later years working as a housekeeper at the Dorchester Hotel, before her death in 1988.

When the film *The Dam Busters* was being shot at Scampton in 1954, the RAF explosives-sniffer dog that played the part of Nigger resolutely declined to approach the spot outside the CO's office where Aircraftman Munro had buried Gibson's beloved black friend on that May night eleven years earlier. For American audiences, it was thought judicious to weave into the film's sequence depicting the raid some images of USAAF B-17 Flying Fortresses.

2 RECONCILIATIONS

Two contending sensations have suffused the author's mind through the creation of this narrative, and a completely satisfactory reconciliation of them seems unattainable. First is that of admiration for the brilliance and personality of Barnes Wallis, together with awe for the achievement of 617 Squadron and especially Guy Gibson, its leader. The other is horror, at the Biblical catastrophe which Operation Chastise unleashed upon those caught in the path of the Möhne and Eder floods. For perhaps two generations after the war, the story of the dambusters raid inspired unqualified triumphalism among the British people. There was little, if any, awareness that many of its victims were foreign captives, friends of the wartime Grand Alliance. There was also a powerful belief that whatever Germans suffered that night, and indeed throughout the war, they had brought upon themselves by becoming first the followers, then the instruments, of Hitler.

Yet Guy Gibson became conscious of the raid's victims. Writing in 1944, he described a conversation through the intercom in G-George on the way back from the Eder. One of his crew suggested that the floods might quench the fires lit by Main Force in Duisburg the previous night – in truth, it was three nights since 572 of Harris's aircraft had attacked the city. 'Someone else said, rather callously I thought, "If you can't burn 'em, drown 'em."' The pilot himself added: 'But we had not tried to do this; we had merely destroyed a legitimate industrial objective so as to hinder the Ruhr Valley output of war munitions. The fact that people were in the way was incidental. The fact that they might drown had not occurred to us … Nobody liked mass slaughter, and we did not like being the authors of it. Besides, it brought us in line with Himmler and his boys.' Somebody – perhaps more than one somebody – must have questioned the morality of the dams raid, in a fashion that lingered in Gibson's mind, troubling him sufficiently that he mentioned it in his memoir, in tones of unease never avowed by his superior officers. This is both fascinating and impressive.

Moral argument about the breaching of the Möhne and the Eder belongs in the same forum as debate about the entire strategic bomber offensive. Chastise cannot be ring-fenced, comfortably distanced from the rest. Although I myself have been for four decades a critic of Harris, the USAAF's Gen. Curtis LeMay – architect of the 1945 destruction of Japan's cities – and their fellow 'bomber barons', I reject the allegation sometimes advanced today, that the wartime bomber offensive should be viewed, with hindsight, as a properly indictable 'war crime'.

Unlike the Holocaust, or German and Japanese massacres of PoWs and civilians of many nationalities, which were carried out in pursuit of deranged ideological, doctrinal and cultural principles, bombing was sincerely intended and expected by its advocates to advance Allied victory; to hasten the destruction of Nazism and bring about the liberation of the tens of millions whom it had enslaved, ironically including the likes of those who perished beneath the Möhne flood. From 1938 to 1943, the RAF's leaders, and later Barnes Wallis, promoted destruction of the dams – as Gibson wrote – because of the damage they thus hoped to inflict upon German industry and infrastructure. Commanders may be accused of callousness because, from their viewpoint on 'the other side of the hill', they closed their minds to the horrors such action must precipitate. But they harboured no direct ambition to precipitate such an outcome.

Sir Michael Howard, the eminent historian who is today among the last survivors of the generation that fought in the war, is a man of supreme humanity. Yet he is also an unyielding defender of the strategic air offensive: 'It is impossible to fight a total war with clean hands. Not merely the Nazis, but the German people, resisted until the bitter end. The Allies were obliged to use every weapon at their disposal to force them to yield.' At the time of the Dresden firestorm in February 1945, Howard observes, the enemy were still killing as many British soldiers as they could contrive; Victoria Crosses were being won for some of the heroic efforts that continued to be necessary to overcome resistance. In the context of Chastise, Howard adds: 'It was no worse to drown Germans than to burn their cities.' This view may seem harsh to some twenty-first-century readers, but the former Coldstream Guards officer is too significant a witness, as a voice for his generation, for us to fail to place his testimony in the balance.

I strive to avoid a vice common among modern writers, of judging events of the past, in vastly different circumstances and a desperate national predicament, by the standards of our own privileged times. Our forefathers, conducting a struggle for survival, made war as they could, rather than as they should, and deserve sympathy for their decisions. Nonetheless, I concluded my 1979 book *Bomber Command*: 'The cost of the bomber offensive in life, treasure, and moral superiority over the enemy tragically outstripped the results that it achieved.' Forty years on, I see no reason to modify that view. It is striking that such a distinguished historian as Richard Overy, who for much of his career was a warm advocate for Bomber Command, recently made a dramatic recantation, albeit not explicitly avowed as such. He wrote in 2013: 'There existed throughout the conflict a wide gap between what was claimed for bombing and what it actually achieved in material and military terms. Just as there existed a wide gap between the legal and ethical claims made on its behalf and the deliberate pursuit of campaigns in which civilian deaths were anticipated and endorsed. The resources devoted to strategic bombing might more usefully have been used in other ways.'

No modern democratic government which dared to undertake such an operation against an enemy as Chastise could retain the support of its own people. Even at the height of the Vietnam war, US president Lyndon Johnson declined to authorise air attacks to breach the dykes and dams of the communist North, which his generals urged him to undertake. Johnson understood that such action must incur the outrage of most Americans, not to mention world opinion. Under the terms of the 1977 Protocols to the Geneva Convention, deliberate precipitation of such catastrophes as the breaching of the Möhne and Eder would constitute a war crime.

1943, however, was not 1966 or 1977. The aircrew who executed Chastise will continue to deserve the admiration of their descendants as long as the British, together with the

Canadian, Australian, New Zealand and American peoples, cherish their wartime heritage. They too, like all the men of Bomber Command, were victims, in that their commanders deliberately obscured from them the commitment of the 1942– 45 offensive to the systematic terrorisation and killing of German civilians, together with their own meagre prospects of survival. Freeman Dyson wrote of his own experience at Harris's headquarters: 'Secrecy pervaded the hierarchy from top to bottom, not so much directed against the Germans, as against the possibility that the failures and falsehoods of the Command should become known either to the political authorities in London or to the boys on the squadrons.' Bomb-aimer Len Sumpter spoke with a moving inarticulacy about 617 Squadron's briefing before Chastise: 'The head people said it would do this and that, put the steel works out of action and what not.'

The outcome of the operation, as of the entire bomber offensive, showed that the warmaking capacity of such a formidable nation as Germany was vast, widely dispersed and resilient. There was never a formula for the destruction from the air of the Nazi industrial edifice with the means then available, whether through bombing cities, synthetic-oil plants, ball-bearing factories, aircraft plants, railways – or dams. The operation demonstrated the wisdom of the warning delivered earlier in the war by the great scientist civil servant Sir Henry Tizard, who grasped air power's limitations with a clarity that eluded Cherwell, Portal, Harris, Spaatz, Eaker and other air chiefs. Tizard wrote to Cherwell in April 1942 that he did not doubt the power of sustained attack to make a *catastrophic* impact on Germany. He questioned, however, whether this would prove *decisive*, as Cherwell and others promised when they embarked upon the air campaign, and especially when they mandated 'area bombing'.

Tizard was proved right. The bomber offensive, and indeed its most spectacular moment at the Möhne, generated a catastrophe, but failed to inflict the crippling blow upon the Nazi industrial machine, the 'disaster of the first magnitude', that Portal and Barnes Wallis had sought and promised. From the perspective of the twenty-first century, we should be grateful that our own society confronts no such peril as was posed by the historic evil of Hitler's Germany, which caused those striving against it to embrace policies that were both ruthless and strategically deluded.

A school friend of Henry Maudslay, the humorist Michael Bentine, said long afterwards of the pilot and many other contemporaries who perished: 'I suddenly see in my mind this boy and think, "Gee, wouldn't it have been marvellous if they had survived, what good they would have done in the world; they would been great doctors, engineers, scientists." I feel very sad when I think of Eton.' And a host of other schools, together with their lost boys and girls.

In the central square of Neheim, the first large community in the path of the Möhne torrent, there stands today a memorial to '1285 Women, Men and Children dead in the flood – 181 citizens of the town, 721 foreign civilians and PoWs'. In Neheim's densely wooded cemetery, where many of the *Möhnekatastrophe*'s victims were interred in mass graves, only the Germans are remembered by names: 'Rita Lübke, Sofia Lübke, Erika Schulte, Margret Schulte, Elisabeth Schulte, Berneide Mette, Maria Mette'. And many more.

A survivor of the deluge, Russian forced labourer Elena Wolkova from Siberia, had for months been conducting a secret affair with a German guard, Karl Josef Stuppardt, who on the fatal night raised the alarm and helped some women workers to flee. Nazi displeasure meant that his relationship with Elena cost him a brief prison term, but on 16 June 1945 the couple were married in the church of St John the Baptist in Neheim, where two years earlier many of the dead from the flood had been laid out. The bride wore her sister-in-law's wedding dress; a choir of liberated Russian prisoners sang in the organ loft.

On the dam itself another memorial, erected in 2015, records in German and English: 'A trail of devastation was left throughout the Möhne valley ... Men, women and children died in the wide torrent of water ... Among the victims were hundreds of female forced labourers from the Ukraine, Poland and Russia ... This strategic action did not have the military effect expected. This memorial is also a symbol of peace.'

Appendix I

617 Squadron's Crews Who Flew on the Night of 16/17 May 1943

(with decorations awarded to survivors for the raid.
† denotes a fatal casualty; * denotes taken as a
prisoner of war)

THE FIRST WAVE

Aircraft	Pilot	Flt Eng.	Navigator	Wireless-Operator	Bomb-Aimer	Front-Gunner	Rear-Gunner
G-George	W/Cdr. Guy Gibson (VC)	Sgt. John Pulford (DFM)	P/O Harlo Taerum (DFC)	F/Lt. Bob Hutchison (bar DFC)	P/O Fred Spafford (DFC)	F/Sgt. George Deering (DFC)	F/Lt. Richard Trevor-Roper (DFC)
M-Mother	F/Lt. John Hopgood †	Sgt. Charles Brennan †	F/O Ken Earnshaw †	Sgt. John Minchin †	P/O John Fraser *	F/O George Gregory †	P/O Tony Burcher *
P-Popsie	F/Lt. Harold Martin (DSO)	P/O Ivan Whittaker	F/Lt. Jack Leggo (bar DFC)	F/O Len Chambers	F/Lt. Bob Hay (bar DFC)	P/O Toby Foxlee	F/Sgt. Tom Simpson (DFM)
A-Apple	S/Ldr Melvin Young †	Sgt. David Horsfall †	F/Sgt. Charles Roberts †	Sgt. Lawrence Nichols †	F/O Vince MacCausland †	Sgt. Gordon Yeo †	Sgt. Wilf Ibbotson †

Aircraft	Pilot	Flt Eng.	Navigator	Wireless-Operator	Bomb-Aimer	Front-Gunner	Rear-Gunner
J-Johnny	F/Lt. David Maltby (DSO)	Sgt. Bill Hatton	Sgt. Vivian Nicholson (DFM)	Sgt. Antony Stone	P/O John Fort (DFC)	Sgt. Vic Hill	Sgt. Harold Simmonds
L-Leather	F/Lt. David Shannon (DSO)	Sgt. Robert Henderson	F/O Danny Walker (bar DFC)	F/O Brian Goodale	F/Sgt. Len Sumpter (DFM)	Sgt. Brian Jagger	P/O Jack Buckley (DFC)
Z-Zebra	S/Ldr. Henry Maudslay †	Sgt. Jack Marriott †	F/O Robert Urquhart †	WO2 Alden Cottam †	P/O Michael Fuller †	F/O Johnny Tytherleigh †	Sgt. Norm Burrows †
B-Baker	Fl/Lt. Bill Astell †	Sgt. Jock Kinnear †	P/O Floyd Wile †	WO2 Abram Garshowitz †	F/O Don Hopkinson †	F/Sgt. Frank Garbas †	Sgt. Richard Bolitho †
N-Nuts	P/O Les Knight (DSO)	Sgt. Ray Grayston	F/O Harold Hobday (DFC)	F/Sgt. Robert Kellow	F/O Edward Johnson (DFC)	Sgt. Fred Sutherland	Sgt. Harry O'Brien

THE SECOND WAVE

Aircraft	Pilot	Flt Eng.	Navigator	Wireless-Operator	Bomb-Aimer	Front-Gunner	Rear-Gunner
E-Easy	F/Lt. Norm Barlow †	P/O Leslie Whillis †	F/O Philip Burgess †	F/O Charles Williams †	P/O Alan Gillespie †	F/O Harvey Glinz †	Sgt. Jack Liddell †
W-Willie	F/Lt. Les Munro	Sgt. Frank Appleby	F/O Jock Rumbles	Sgt. Percy Pigeon	Sgt. Jim Clay	Sgt. Bill Howarth	F/Sgt. Harvey Weeks
K-King	P/O Vernon Byers †	Sgt. Alastair Taylor †	F/O Jim Warner †	Sgt. John Wilkinson †	P/O Neville Whitaker †	Sgt. Charles Jarvie †	F/Sgt. Jim McDowell
H-Harry	P/O Geoff Rice	Sgt. Edward Smith	F/O Richard MacFarlane	Sgt. Chester Gowrie	F/Sgt. John Thrasher	Sgt. Tom Maynard	Sgt. Steve Burns
T-Tommy	F/Lt. Joe McCarthy (DSO)	Sgt. Bill Radcliffe	F/Sgt. Don MacLean (DFM)	Sgt. Len Eaton	Sgt. George Johnson (DFM)	Sgt. Ron Batson	F/O David Rodger

THE RESERVE WAVE

Aircraft	Pilot	Flt Eng.	Navigator	Wireless-Operator	Bomb-Aimer	Front-Gunner	Rear-Gunner
C-Charlie	P/O Bill Ottley †	Sgt. Ron Marsden †	F/O Jack Barrett †	Sgt. Jack Guterman †	F/Sgt. Tom Johnston †	Sgt. Harry Strange †	Sgt. Fred Tees
S-Sugar	P/O Lewis Burpee †	Sgt. Guy Pegler †	Sgt. Tom Jaye †	P/O Len Weller †	WO2 Jim Arthur †	Sgt. Bill Long †	WO2 Joe Brady †
F-Freddie	F/Sgt. Ken Brown (CGM)	Sgt. Basil Feneron	Sgt. Dudley Heal (DFM)	Sgt. Herb Hewstone	Sgt. Stefan Oancia (DFM)	Sgt. Dan Allaston	F/Sgt. Grant McDonald
O-Orange	P/O Bill Townsend (CGM)	Sgt. Dennis Powell	P/O Lance Howard (DFC)	F/Sgt. George Chalmers (DFC)	Sgt. Charles Franklin (bar DFM)	Sgt. Doug Webb (DFM)	Sgt. Ray Wilkinson (DFM)
Y-York	F/Sgt. Cyril Anderson	Sgt. Robert Paterson	Sgt. John Nugent	Sgt. Doug Bickle	Sgt. Gilbert Green	Sgt. Eric Ewan	Sgt. Arthur Buck

Key to photographs of Chastise aircrew:

(Row 1, left to right) **G-George:** Gibson, Pulford, Taerum, Hutchison, Spafford, Deering, Trevor-Roper

(Row 2) **M-Mother:** Hopgood, Brennan, Earnshaw, Minchin, Fraser, Gregory, Burcher

(Row 3) **P-Popsie:** Martin, Whittaker, Leggo, Chambers, Hay, Foxlee, Simpson

(Row 4) **A-Apple:** Young, Horsfall, Roberts, Nichols, MacCausland, Yeo, Ibbotson

(Row 5) **J-Johnny:** Maltby, Hatton, Nicholson, Stone, Fort, Hill, Simmonds

(Row 6) **L-Leather:** Shannon, Henderson, Walker, Goodale, Sumpter, Jagger, Buckley

(Row 7) **Z-Zebra:** Maudslay, Marriott, Urquhart, Cottam, Fuller, Tytherleigh, Burrows

(Row 8) **B-Baker:** Astell, Kinnear, Wile, Garshowitz, Hopkinson, Garbas, Bolitho

(Row 9) **N-Nuts:** Knight, Grayston, Hobday, Kellow, Johnson, Sutherland, O'Brien

(Row 10) **E-Easy:** Barlow, Whillis, Burgess, Williams, Gillespie, Glinz, Liddell

(Row 11) **W-Willie:** Munro, Appleby, Rumbles, Pigeon, Clay, Howarth, Weeks

(Row 12) **K-King:** Byers, Taylor, Warner, Wilkinson, Whitaker, Jarvie, McDowell

(Row 13) **H-Harry:** Rice, Smith, MacFarlane, Gowrie, Thrasher, Maynard, Burns

(Row 14) **T-Tommy:** McCarthy, Radcliffe, MacLean, Eaton, Johnson, Batson, Rodger

(Row 15) **C-Charlie:** Ottley, Marsden, Barrett, Guterman, Johnston, Strange, Tees

(Row 16) **S-Sugar:** Burpee, Pegler, Jaye, Weller, Arthur, Long, Brady

(Row 17) **F-Freddie:** Brown, Feneron, Heal, Hewstone, Oancia, Allatson, McDonald

(Row 18) **O-Orange:** Townsend, Powell, Howard, Chalmers, Franklin, Webb, Wilkinson

(Row 19) **Y-York:** Anderson, Paterson, Nugent, Bickle, Green, Ewan, Buck

Appendix II

Landmark Dates in the Evolution of Chastise

Oct 1937 – RAF planners identify German dams as important strategic targets

Feb 1940 – Barnes Wallis works for Vickers on advanced Wellington bomber, moonlights on his own 'Victory' bomber and giant 'earthquake' bomb projects

June 1940 – Gp. Capt. Norman Bottomley of RAF Bomber Command proposes attack on dams

July 1940 – Sir Charles Portal, then Bomber Command C-in-C, urges attack on the Möhne

Nov 1940 – Wallis shown secret Air Ministry studies on dams; Road Research Laboratory begins construction of scaled-down models for Ministry of Aircraft Production; Wallis continues work on his 'earthquake' bomb

March 1941 – Wallis circulates paper, 'Note on Methods of Attacking Dams', using big bombs

Air Ministry forms AAD – Aerial Attack on Dams – Committee, which hears of 'earthquake' bomb

21 May 1941 – Sir Henry Tizard tells Wallis that his proposals for a Victory bomber and 'earthquake' bomb have been rejected by AAD Committee. Research nonetheless continues on ballistics of destroying dams

Feb–March 1942 – Arthur Collins of the Road Research
Laboratory conducts experiments with contact charges
against model dams at Harmondsworth which prompt …

April 1942 – … Wallis to make experiments at Effingham
which convince him of the feasibility of attacking large
structures or ships across water, by bouncing bombs

May 1942 – Wallis publishes for secret circulation a paper,
'Spherical Bomb, Surface Torpedo', explaining his
theories

May–July 1942 – Collins of the RRL conducts further
experiments at Harmondsworth and Nant-y-Gro which
conclude that a dam might be broken with a relatively
small charge, delivered directly against the masonry

June–Sept 1942 – Wallis conducts tests at Teddington ship
tanks, with support from the Royal Navy. Admiralty
remains enthusiastic about Highball, an anti-ship
bouncing bomb, while MAP question economics of
modifying Lancasters to carry Upkeep against dams

22 July 1942 – MAP hesitantly authorises modification of a
Wellington to carry dummy Upkeeps, orders twelve test
examples of bombs

4 Dec 1942 – First aerial tests at Chesil Beach, Dorset

15 Dec 1942 – Second Chesil Beach trials

9 Jan–24 Feb 1943 – Further intermittent Chesil Beach trials
with modified Upkeeps

12–13 Feb 1943 – Wallis informed that MAP considers
Upkeep project unacceptably speculative, while Air
Ministry questions feasibility of aircrew delivering bombs
to dams. Harris of Bomber Command characterises
proposals as 'tripe'

15 Feb 1943 – At Air Ministry meeting, Wallis persuades
Bottomley and Bufton of Upkeep's feasibility

19 Feb 1943 – Portal tells Harris he is minded to support Wallis's bombs; three Lancasters to be modified for tests

22 Feb 1943 – Wallis meets Harris at High Wycombe

23 Feb 1943 – Vickers chairman Sir Charles Craven orders Wallis to abandon bouncing bombs because of MAP's loss of confidence in project. Wallis threatens resignation

26 Feb 1943 – Wallis attends meeting at MAP at which he is informed that Portal has committed himself to launching a squadron-sized operation against Germany's dams with Upkeep

27 Feb 1943 – First Sea Lord declares unequivocal support for Highball project. Sixteen Coastal Command Mosquitoes earmarked to carry the bombs

28 Feb 1943 – Drawings of spherical Upkeeps dispatched by Wallis to Vickers

6/7 March 1943 – Harris launches Main Force 'Battle of the Ruhr'

8 March 1943 – First Type 464 'Provisioning' Lancaster ordered

18 March 1943 – Guy Gibson meets 5 Group AOC, ordered to form 'Squadron X' to attack unidentified low-level targets

23 March 1943 – 617 Squadron officially formed

27 March 1943 – Portal submits paper to chiefs of staff explaining Chastise

1 April 1943 – Chiefs of staff informed that only twenty Lancasters – in the event twenty-three – would be built to carry Upkeep

13 April 1943 – Upkeep trials at Reculver fail

18 April 1943 – Upkeep trials at Reculver fail. Decision to remove wooden casing, and instead work with steel cylinder

21 April 1943 – Upkeep trials at Reculver fail

24 April 1943 – Wallis tells Gibson Chastise doomed unless 617 Squadron can attack at a height of sixty feet

29 April 1943 – Gibson witnesses successful drop of Upkeep at Reculver from sixty feet

1 May 1943 – Gibson informs Wallis his crews can attack at sixty feet

4 May 1943 – Gibson informs 5 Group his crews ready to operate

9/10 May 1943 – Highball trials at Loch Striven fail [and likewise later on 17/18 and 22/23 May]

11/12 May 1943 – 617 crews drop dummy Upkeeps at Reculver

13 May 1943 – Gibson witnesses drop and detonation of live Upkeep off Broadstairs

14 May 1943 – 617 Squadron conducts 'dress rehearsal' at Eyebrook and Abberton reservoirs

15 May 1943 – Air Ministry sends Bomber Command order to execute Chastise 'at first suitable opportunity'

Appendix III

A Chronology of Operation Chastise 16/17 May 1943

Sunday, 16 May 1943

2100 – Red Very light fired for First and Second Waves to start engines

2128 – Barlow takes off in E-Easy

2129 – Munro takes off in W-Willie

2130 – Byers takes off in K-King

2131 – Rice takes off in H-Harry

2139 – Gibson in G-George, Martin in P-Popsie, Hopgood in M-Mother take off

2147 – Young in A-Apple, Maltby in J-Johnny, Shannon in L-Leather take off

2154 – Munro crosses English coast

2159 – Maudslay in Z-Zebra, Astell in B-Baker, Knight in N-Nuts take off

2201 – McCarthy takes off in T-Tommy

2229 – Gibson section crosses English coast

2238 – Young section crosses English coast

2247 – Maudslay section crosses English coast

2256 – Munro reaches Dutch coast

2257 – Munro hit

2257 – Byers shot down

2259 – Rice crosses Dutch coast

2306 – Rice hits water, turns back

2300 – Wallis and Cochrane leave Scampton for Grantham

2302 – Gibson, Hopgood, Martin cross Dutch coast

2306 – Munro turns for home

2312 – Young, Maltby, Shannon reach Dutch coast

2313 – McCarthy reaches Vlieland

2321 – Maudslay trio cross Scheldt

2322 – Maudslay, Astell, Knight reach Dutch coast

2350 – Barlow crashes

Monday, 17 May

0007 – Gibson trio encounters flak – Hopgood's plane apparently hit. Gibson transmits flak warning

0009 – Ottley takes off in C-Charlie

0011 – Burpee takes off in S-Sugar

0012 – Brown takes off in F-Freddie

0014 – Townsend takes off in O-Orange

0015 – Anderson takes off in Y-York, Gibson, Martin, Hopgood reach Möhne – Martin first, Astell hits high-tension cable, McCarthy reaches Sorpe

0016 – Munro crosses back over English coast

0026 – Young trio reach Möhne, Shannon arrives late

0028 – Gibson attacks

0033 – Hopgood attacks and is shot down

0036 – Munro lands back at Scampton

0037 – Gibson's first 'GONER68A' sent

0038 – Martin attacks

0043 – Young attacks

0046 – McCarthy drops Upkeep at Sorpe

0047 – Rice lands back at Scampton

0048 – Young's 'GONER78A' signal received at Grantham

0049 – Maltby attacks Möhne

0050 – 'GONER78A' signalled from Maltby, though dam already crumbling

0053 – Martin's 'GONER58A' signal received

0056 – Gibson orders 'NIGGER' signal to be sent

0130 – Gibson arrives at Eder and fires Very light, Brown reaches Dutch coast

0131 – Townsend reaches Dutch coast

0139 – Shannon attacks Eder

0145 – Maudslay attacks Eder, Townsend evades enemy flak

0146 – 'GONER28B' sent

0147 – Gibson calls up Astell

0150 – Gibson again calls up Astell, Knight attacks Eder

0153 – Maltby crosses back over Dutch coast, Burpee shot down

0154 – 'DINGHY' sent – Eder breached

0155 – Grantham asks for confirmation

0157 – Maudslay sends signal to Grantham

0206 – Shannon's 'GONER79B' message received at Grantham

0210 – Grantham asks Gibson if any First Wave Upkeeps available for Sorpe

0222 – Townsend ordered to attack Ennepe

0228 – Anderson ordered to attack Sorpe

0230 – Grantham tries to divert Burpee to Sorpe

0231 – Ottley instructed to attack Lister, signal acknowledged

0232 – Ottley redirected to Sorpe

0235 – Ottley crashes near Hamm

0236 – Maudslay shot down

0258 – Young shot down

0259 – Knight crosses Dutch coast

0310 – Anderson turns for home

0311 – Maltby lands

0314 – Brown attacks Sorpe

0319 – Martin lands

0323 – 'GONER78C' from Brown, McCarthy lands

0337 – Townsend attacks Ennepe

0400 – Wallis, Harris, Cochrane leave Grantham for Scampton

0406 – Shannon lands

0415 – Gibson lands

0530 – Anderson lands

0533 – Brown lands

0615 – Townsend lands

Acknowledgements

My foremost debt in the writing of this book is to witnesses who are now long dead: chiefs and aircrew of Bomber Command, whom I had the privilege of meeting back in 1977–78, when I was researching my history of the strategic offensive. I learned much from them which has influenced this narrative, both about wartime decision-making and about the experience of flying heavy bombers over Germany. I should also pay tribute to Michael Anderson's 1955 movie *The Dam Busters*, which possesses a peerless period feel, enhanced by the fact that many of those who acted in it had themselves served in the Second World War. Few, if any, twenty-first-century players can match the spirit – the curious and moving innocence and gravity – displayed by that past generation of actors and actresses in depicting 1943 fliers, and indeed people of all sorts who experienced the period.

In our own times I am indebted to Richard Morris, author of authoritative biographies of Guy Gibson and Leonard Cheshire, and shortly to publish a new study of Barnes Wallis, which will contribute importantly to our understanding of that remarkable man's life and work. Many writers about the Second World War, and especially about its air campaigns, focus either upon heroes or technical detail, heedless of the 'big picture'. Richard, by contrast, is a penetrating student of both character and the significance of events.

Robert Owen, historian of 617 Squadron and author of a 2015 biography of Henry Maudslay, is a fount of information about every detail of Chastise. Again and again, when I have consulted him about events, timings, personnel, he has responded exhaustively and generously. Where facts and timings are disputed, I have no hesitation in accepting his verdicts, and am boundlessly grateful for both his counsel and his company, especially on my own 2018 visit to the dams. Neither Rob nor Richard, of course, bears any responsibility for my errors, nor shares complicity in my judgements. My old friend and colleague Patrick Bishop, himself a distinguished author, generously gave me access to files he researched for his own superb books on the wartime RAF, especially those from the Brotherton Library, Leeds. Richard Mead, who is writing a biography of Ralph Cochrane, kindly answered a couple of questions about that somewhat opaque, though indisputably able officer. In May 2018 the Royal Aeronautical Society hosted a conference to mark the seventy-fifth anniversary of Chastise, at which a succession of speakers provided provocative and helpful reflections on the operation.

My old friend Professor Sir Michael Howard, OM, CH, MC, has once again read my manuscript and made some penetrating and important comments and observations, especially valuable because he is among the last survivors of the generation that was there – that fought in the Second World War.

In Germany, Werner Buehner spent two days driving me and Rob Owen around the dams, explaining a host of details that could only have been mastered by a man who has devoted a lifetime of his leisure hours to exploring the subject. Above the Möhne, local farmer Jochen Peters is custodian of the crash site of John Hopgood's M-Mother, and spent two hours describing local landmarks.

The librarians at Christ Church College, Oxford, assisted me with access to the Portal Papers; Seb Cox and his staff made me welcome in the Air Historical Branch archive, now located at RAF Northolt; Peter Devitt, archivist of the RAF Museum, enabled me to read the unpublished, annotated typescript of Guy Gibson's *Enemy Coast Ahead*. I have been researching, often happily and always profitably, at the Imperial War Museum on and off for well over half a century. The National Archives at Kew are among Britain's great institutions, a scholarly treasure trove. Will Spencer's last kindness to me before his retirement from the NA, after decades of providing assistance to such historians as myself, was to arrange access to files relating to Chastise. These must nowadays, alas, be read under secure conditions, following abuses by mischievous or recklessly irresponsible 617 Squadron obsessives. The London Library remains an indispensable, as well as joyous, rendezvous for every researcher.

For many years now I have enjoyed with Arabella Pike, Robert Lacey, Helen Ellis and their colleagues at HarperCollins the happiest relationships I have ever achieved with publishers: I remain boundlessly grateful for their contributions to my books. Every day I find myself lamenting the 2018 death of Michael Sissons, for almost forty years a close friend and counsellor, as well as my literary agent: *Chastise* is the last book for which he negotiated the British contract. It is a great sadness that he did not live to read it, though I warmly welcome the support of Andrew Wylie, who now represents me.

My personal assistant, Rachel Lawrence, gives me as much priceless aid in 2019 as she did when she first fulfilled this role in my life, at the *Daily Telegraph*, back in 1986. My wife Penny remains the light of my life, as well as the most astringent and just critic of both my follies and my writing. Without her, the former would have long since beached me; the latter would never happen.

Notes and References

'AI' denotes an interview conducted by the author in 1977–78, for *Bomber Command*. 'ECA' signifies a quotation from either the unpublished typescript of Guy Gibson's *Enemy Coast Ahead*, held at the RAF Museum in Hendon, or from the published version, as respectively indicated. 'UKNA' abbreviates Britain's National Archives at Kew. 'USNA' denotes America's National Archives. 'IWM' indicates the documentary or oral archive of the Imperial War Museum. 'CCh College' signifies Christ Church College, Oxford, where the Portal Papers are held. No references are given for quotations from statements and documents that have been long in the public domain, and are widely cited elsewhere.

Epigraphs
  xv 'It is proposed' Oxland in UKNA AIR14/840
  xv 'One thing' Gibson *ECA* p.203. The reported exchange concerned a prospective July 1942 attack on German capital ships

Introduction
xxi 'The story of 617 Squadron's' Morris, Richard with Colin Dobinson *Guy Gibson* Viking 1994 p.184

xxii   'The fact that' Gibson, Guy *Enemy Coast Ahead* unpublished draft p.298

xxiii   'sanctimonious, hypocritical' IWM Oral Archive Shannon 16256/2 reel 2

Prologue

xxix   'The moon was full' Gibson, Guy *Enemy Coast Ahead* Michael Joseph 1946 p.17

xxix   'full of dull' ibid. p.2

xxx   'We were off' ibid. pp.17–18

xxx   'a mission which' Gibson letter to Pearn, Pollinger & Higham 1.8.44

xxxi   'The pilot sits' Gibson *ECA* pp.17–18

xxxii   'two silent figures' ibid. p.18

xxxiv   'He was the sort' AI Bomber Command veteran 1977

xxxiv   'Our noses were' Gibson *ECA* p.21

xxxiv   'One hour to go' ibid. unpublished draft p.24

xxxv   'it's exciting' *Desert Island Discs* 19.2.44

xxxv   'This was the big thing' Gibson *ECA* unpublished draft p.4

1: Grand Strategy, Great Dams

2   'All the old animosities' USNA OWI surveys Nos 113, 114, 117

3   'the feeling that' Elmes, Jenny *M-Mother: Dambuster John 'Hoppy' Hopgood* The History Press 2015 p.178

3   'It was pleasant' Gibson *ECA* unpublished draft p.301

5   'There must be action' Hastings, Max *Finest Years: Churchill as Warlord 1940–45* HarperCollins 2009 p.114

8   'According to him' Alanbrooke, Lord *War Diaries 1939–45* ed. Danchev & Todman Weidenfeld & Nicolson 2001 13.10.43 p.460

9 'typical example' Dyson, Freeman *Disturbing the Universe* Harper & Row 1979 p.29

9 'There was a certain' Brown, Anthony Montague *Long Sunset: The Memoirs of Churchill's Last Private Secretary* Cassell 1995 p.201

10 'He experienced' UKNA AIR2/5995

10 'I admired' Montague Brown p.201

11 'It's very nice' Probert, Henry *Bomber Harris: His Life and Times* Greenhill 2006 p.73, and elsewhere in his pages for a frank discussion of Harris's complicated personal life

12 'in their efforts' Hastings, *Bomber Command* p.178

13 'And what have you' AI Bufton 1978

14 'It does not appear' CCh College Portal Papers File 10

14 'I feel bound to tell you frankly' ibid. Box 10 10.10.42

17 'while the RAF was doctrinally committed' AI Wallis 1977

18 'the weight of attack will invariably fall' Slessor, John *Air Power and Armies* London 1936 p.65

19 'enormous damage' Sweetman, John *The Dambusters Raid* Arms & Armour Press 1993 p.3

19 'again urged the merit' Almost the whole history of the RAF's wartime engagement with dams, and especially that of Barnes Wallis, is to be found in the fascinating documentation contained in the British National Archives file UKNA AVIA53/627

20 'comparison of the Results' Air Historical Branch Monograph *Armament* Vol. I 1952 p.17

20 'Give us the tools' WSC February 1941 broadcast to the US

22 'there seems a probability' Sweetman p.8

2: The Boffin and His Bombs

24 'the "Dam Buster" AHB *Armament* Vol. I p.222

27 'My dear boy' Winterbotham *Secret and Personal* p.156

27 'We were used to' Morris, Richard ed. *Breaching the German Dams: Flying Into History* RAF Museum 2008 p.11

28 'He was a collision' Richard Morris to the author March 2019

29 'All my heart is in airships' Morpurgo, J.E. *Barnes Wallis* Longman 1972 p.102

30 'Cherwell likewise promoted a CS' AHB *Armament* Vol. I p.19

30 'There were many' Boorer quoted in Arthur, Max *Dambusters: A Landmark Oral History* Virgin Books 2009 p.46

31 'Life is almost unrelieved' Morpurgo p.224 letter to Masterman 1.8.40

31 'is going to be the instrument' specification of 19.7.40, Vickers Papers, 780, Cambridge University Library

33 'From October onwards' AHB monograph *Armament* Vol. I p.19

34 'Wallis was granted access' UKNA AVIA53/627

35 'while their dam' information from BRE website

36 'Everything is being done' WSC note dispatched on 7.10.41, CCh College Portal Papers Folder 2, File 2

38 'Sir, I have the honour' UKNA AIR14/840

39 'the inception of' UKNA AVIA53/627

40 'later claimed credit' Richard Morris to the author March 2019

42 'rota-mines' Winterbotham op. cit. p.148

43 'A solution to' Holland, James *Dam Busters* Corgi 2013 p.106

44 'It is quite' ibid. p.113

45 'If this new weapon' Winterbotham op. cit. p.152

46 'I do wish you could' Morris ed. *Breaching the German Dams* p.9

49 'We have just' UKNA AIR14/840

51 'It was agreed' Holland p.146

51 'Nobody could play' AI Bufton 1977

53 'The potential value' UKNA ADM186/414

54 'your scepticism' Elworthy quoted Probert p.254

54 'What is it' Morpurgo p.248

57 'The MAP's budget' UKNA AVIA53/627

3: Command and Controversy

60 'The former is' UKNA AIR14/840

61 'the speed with which' quoted Holland p.216

63 'two weeks would provide' UKNA AIR14/840

64 'Young man' AI Cochrane 1978

65 'a ruthless martinet' Frankland, Noble *History at War* DLM 1998 p.23

65 'Group headquarters are funny' Gibson *ECA* p.238

67 'erratic, impulsive' information from Richard Morris 2018

68 'the sort of boy' AI Cochrane 1978

69 'I was no serviceman' Gibson *ECA* unpublished draft p.12

69 'Now's your chance' Morris *Guy Gibson* p.37

69 'As usual in these' Gibson *ECA* unpublished draft Chap. 3 p.106

70 'I was still smarting' ibid. p.55

70 'This was due to' ibid. p.130

71 'I was getting' ibid. Chap. 7 p.147

71 'Don't bother about' ibid. p.159

72 'promised him a squadron' information from Richard Morris, March 2019

72 'to find her huddled' Gibson *ECA* p.131

72 'Night fighting was all' ibid. p.164

72 'the other thing' ibid. p.165

73 'I'm afraid' ibid. p.176

73 'the arch-bastard' Foster, Charles *The Complete Dambusters: The 133 Men Who Flew on the Dams Raid* History Press 2018 p.65

73 'Neither, I am afraid' *ECA* p.176

74 '"But surely you don't' ibid. p.213

76 'addicted to stress' Morris *Guy Gibson* p.123

77 'These young men' Beevor, Antony *Arnhem* Penguin 2018 p.289

77 'the war has made me old' ibid. p.290

78 'I never really knew' Eve Gibson typescript memoir quoted in Driesschen, Jan van den *We Will Remember Them* Erskine Press, Quidenham, Norfolk 2004 p.29

78 'There was something' Hastings, Max *Warriors* HarperCollins 2005 p.187

84 'The operation against' UKNA AIR14/840

84 'as I always thought' ibid.

85 'The Economic and Moral Effects' ibid.

4: Men and Machines

91 'a crushing disappointment' Frankland p.15

92 '[Gibson] asked me' Birrell, Dave *Big Joe McCarthy* Wingleader 2012 p.87

92 'It was a name transfer' Arthur p.10

93 'a dependable average' Owen, Robert with Steve Darlow, Sean Feast, Arthur Thorning *Dam Busters: Failed to Return* Fighting High Press 2013 p.21

93 'Had our interview' Press *All My Life* pp.19–22

93 'leapt at the opportunity' IWM Oral History Transcript
Shannon 16256/2 reel one

94 'are not getting' Hastings *Bomber Command* p.215

94 'the very poor type' CCh College Portal Papers Box 9
10.1.43

95 'Just a year yesterday' Holland p.280

96 'This member is not' Thorning p.31

97 'Since we are to have' ibid. p.39

97 'Anyone like a game' Cheshire, Leonard *Bomber Pilot*
Hutchinson 1943 p.52

98 'One of the aviators' Thorning p.57

98 'older and worn' ibid. p.66

99 'and haven't I got' ibid. p.97

100 'My Wing Commander here' Owen, Robert *Henry
Maudslay, Dam Buster* Fighting High Press 2014 p.248

101 'courteous, diffident' Slessor, Jack *The Central Blue*
Cassell 1956 p.376

101 'so the result is' Owen *Maudslay* p.217

102 'that seems to be' Elmes p.56

102 'he is incredibly good and kind' ibid. p.188

103 'I'm all against' letter of 16.7.42 ibid. p.170

103 'The theme was that' 2.1.38 ibid. p.25

103 'He looked pale' ibid. p.218

104 'as big a headache' Frankland p.19

105 'Aircrew are becoming' Hastings, *Bomber Command*
p.215

106 'which is much better' Elmes p.221

106 'they wouldn't have had' AI Martin 1978

106 'mad as a grasshopper' Foster *Complete Dambusters* p.134

107 'I loved the life' IWM Sound Archive Hobday
N7298/04/01 p.2

108 'rather hard cases' Munro quoted Arthur p.24

109 'We lived supremely' Lewis, Cecil *Sagittarius Rising* Peter Owen 1936 p.187

109 'I don't know' *Dambusters Remembered* 1983 p.15

109 'One of the reasons' Elmes p.17

112 'You had more comradeship' Chalmers quoted Arthur p.14

113 'I've never met' Trevor-Roper, Hugh *Journals* Tauris 2012 p.88

113 'a very sinister' Humphries, Harry *Living with Heroes: The Dam Busters* Erskine Press 2008 p.10

114 'When we were under' Frankland p.31

115 'Wallis said: "I'm glad' Gibson *ECA* p.249

117 'He said, "If I' IWM Oral History Archive Sumpter 7372/04/02 p.36

118 'We'd work between' ibid. p.23

118 'Thrapston just coming up' ibid. p.7

118 'I hope we didn't scare you' aircrewremembered.com/byers-vernon html

118 'one had a strong urge' IWM Oral History Transcript Shannon 16256/2 reel 2

119 'It's like riding' Arthur p.105

119 'There was one particular' ibid. p.82

119 'another "tour"' Foster *Complete Dambusters* p.76

120 'with whom he had' ibid. p.66

120 'You said "Good morning"' IWM Sound Archive Transcript Sumpter 7372/04/01 p.13

121 'The "kneeling behind' Gibson *ECA* draft Chap. 12

122 'We took it all in good part' Arthur p.81

123 'lack of awareness' Hillary speech of 22.10.42 published in *The Spectator* 15.1.43

123 'prided himself on' AI Martin 1978

124 'he expected to be' Thorning p.121

125 'irritated by his' Fred Sutherland quoted Thorning p.118

125 'He is not so bad' ibid. p.122

125 'You bloody fool' Humphries p.4

126 'one of the finest' IWM Sound Archive Transcript Shannon 8177/3/1 p.4

5: The Brink of Battle

128 'Christ!' Gibson *ECA* p.253

128 'It's amazing to me' Arthur p.82

128 'smashing the Perspex' Foster *Complete Dambusters* p.59

128 'I am now at' ibid. p.59 letter of 13.4.43

128 'It's no use' Gibson *ECA* p.260

129 'It was when you'd' IWM Sound Archive Transcript Sumpter 7372/2/2 p.23

131 'One thought that' IWM Sound Archive Transcript Shannon 8177/3/1 p.10

133 'The foaming water' Gibson *ECA* p.266

133 'In our secret hangar' ibid. p.265

134 'No further trials' UKNA AIR14/840

137 'What price' IWM 86/15/1 Charles Roland Williams Papers 18/4/43

139 'Bags of security' Sweetman p.70

140 'He made you feel' IWM Sound Archive Transcript Hobday 7298/04/01 p.6

140 'Those in the secret' Gibson *ECA* p.269

142 '[The training] seemed' IWM Sound Archive Transcript Shannon 8177/3/2 p.15

142 'I began to get ill' Gibson *ECA* p.271

142 'Are you happy?' Morris *Gibson* p.159

145 'My dear Mummy' 10.5.43, Owen *Maudslay* p.262

146 'It just didn't seem possible' Arthur p.122

146 'Henry Maudslay and his crew' Owen *Maudslay* p.263

147 'that empty noise' Gibson *ECA* p.32

147 'When you see' IWM Sound Archive Transcript Sumpter
　　7372/04/01 p.5

149 'grand chaps' Elmes p.226

150 'I have not the foggiest' quoted Sweetman pp.74–5

150 'Bloody good show' Fay Gillon MS quoted Morris *Gibson*
　　p.162

6: Chastise

156 'Look here, Guy' Gibson *ECA* p.273

156 'And so I went back' ibid. p.276

157 'Just as I said' Humphries pp.1–3

159 'To push the idea' Dyson p.22

160 'That sounded' Johnson quoted Arthur p.136

161 'Main Force' IWM Sound Archive Transcript Hobday
　　7298/04/1 p.7

161 'We weren't too happy' Holland p.385

163 'I immediately thought' Sutherland *Times* obituary
　　24.1.19

163 'I will have ended' Foster, Charles *Breaking the Dams:*
　　The Story of David Maltby and His Crew Pen & Sword
　　2008 p.174

164 'has shown me' Mick Scott in Blythe, Ronald ed. *Private*
　　Words Viking 1991 p.297

164 'I still maintain' Rosewarne May 1940 in ibid. p.30

165 'rather tousled' Gibson *ECA* p.276

166 'We were very impressed' IWM Sound Archive
　　Transcript Hobday 7298/04/01

167 'I realise how difficult' Cochrane quoted Arthur p.143

168 'Your stomach feels' Gibson *ECA* p.189

171 'two Lancasters' UKNA AIR22/129

172 'I had visions' quoted Sweetman p.135

173 'He may make a mess' Humphries p.11

173 'with this enormous thing' IWM Sound Archive Heal 13248/01

183 'Cheerio for now' IWM Williams Papers

7: At the Dams

188 'It looked squat' Gibson *ECA* p.286

190 'The last rashly' IWM Sound Archive Transcript Sumpter 7372/04/4 p.39

190 'The worst time was' ibid. p.44

192 'We were perhaps' ibid. p.45

194 'a huge bloody spurt' IWM Sound Archive Transcript Shannon 8177/3/3 p.27

195 'whereas half the fliers' Dyson p.28

197 'slightly more ham' Gibson *ECA* draft p.8

199 'There was a certain amount' IWM Sound Archive Transcript Shannon 8177/3/3 p.25

200 'Wizard show' AHB files: In a 30.8.88 letter to RAF historian Henry Probert

202 'the man who was' Gibson *ECA* unpublished draft p.417

205 'Josef Ketting' Euler, Helmuth *The Dambuster Raid: A German View* Pen & Sword 2015 p.93

207 'When the mine exploded' IWM Sound Archive George Johnson 17970/01

209 'They didn't know' IWM Sound Archive Sumpter 7372/04/1

211 'To exit from the Eder' IWM Sound Archive Transcript Shannon 16256/2/1

212 'Henry, that's very' Owen *Maudslay* pp.279–80

213 'Dave made a good' Gibson *ECA* p.295

215 'He was the coolest' Foster *Complete Dambusters* p.196

215 'As we came in' IWM Sound Archive Transcript Hobday 7298/04/1 p.25

216 'It was still intact' Kellow, Robert *Paths to Freedom* Kellow 1992 p.xx

216 'a great joy' Sweetman p.120

216 'We could see' IWM Sound Archive Hobday 7298/01

217 'he had taken a stroll' Euler p.189

218 'Good show, boys' Gibson quoted Sweetman p.120

219 'There was nothing wrong' Cochrane in a letter of 4 June 1972

220 'We flew over this' IWM Sound Archive Heal 13248/01

221 'the wreck would have' Euler p.103

222 'Apart from provision' Owen *Maudslay* p.302

222 'My "Gethsemane"' Foster *Complete Dambusters* p.250

223 'This is a Tommy' Euler p.98

225 'We had done' IWM Sound Archive Heal 13248/01

227 'I don't think we ever gave a thought' Arthur p.224

227 'I remember looking' quoted Sweetman p.143

229 'defying the usual' IWM Sound Archive Transcript Sumpter 7372/04/2 p.25

229 'I think we've got to' ibid. p.46

230 'the calm and silvery' Gibson *ECA* p.302

232 'I can still smell' Slessor *Central Blue* p.376

234 'Melvin was apt' Thorning p.92

234 'We were used to it' IWM Sound Archive Transcript Hobday 7298/04/1 p.27

235 'could prove anything really' IWM Sound Archive Transcript Sumpter 7372/04/04 p.42

8: The *Möhnekatastrophe*

236 'wasted precious minutes' Air Historical Branch Translation No. VII/36 The Westphalian President's report to the Minister of Home Affairs 24.6.43

237 'peculiar mushrooms' Quotations from testimony below are taken from Euler, Helmuth *The Dambusters Raid: A German View* unless otherwise stated

239 'when the entire local' Arnsberg *Regierungspresident*'s report, AHB Translation VII/36

240 '170,000 perished' Harvey, Elizabeth *Last Resort or Key Resource? Women Workers from the Nazi-occupied Soviet Territories, the Reich Labour Administration and the German War* Transactions of the Royal Historical Society pp.149–73, 2016 p.151

241 'because of the panic' ibid. p159

241 'We were fed' Euler p.158

243 'Ukrainian Darja Moros' in a letter to Euler of 18.7.90 p.157

244 'their report records' Arnsberg *Regierungspresident* to Minister of Home Affairs 24.6.43

250 'The rescue and identification' 24.6.43 AHB Translation VII/36

255 'describes the impact' *History of the Second World War* Potsdam Vol. IX/1 Oxford 1988 p.381

257 'a mere nineteen' Speer *Inside the Third Reich* pp.280–1

257 'That night [of 16/17 May]' ibid.

258 'an outbreak of panic' UKNA HW16/9

259 'dark as the picture of destruction may seem' AHB Translation VII/36

259 'The wicked little angels' Moltke, Helmuth von *Letters to Freya* Collins Harvill 1988 p.302 17.5.43

9: Heroes

262 'Apart from clothes' National Archives of Canada

266 'The RAF attack ... appears' Lascelles, Sir Alan *King's Counsellor* ed. Duff Hart Davis Weidenfeld & Nicolson 2006 18.5.43 p.130

267 'Comments (all made by men)' Garfield, Simon ed. *Private Battles* Ebury Press 2006 p.356

267 'I don't like it' Koa Wing, Sandra ed. *Our Longest Days* Profile Books 2007 p.179

269 'The last few days' Klemperer, Victor *To the Bitter End* Phoenix 2000 p.281

270 'I write in haste' Morris *Breaching the German Dams* p.55

271 'no major operations' UKNA CAB120/83

271 'realised how near' Alanbrooke *Diaries* pp.409–11 24/25.5.43

272 'Another very disappointing day' ibid. p.405

272 'Portal's official biography' Richards, Denis *Portal of Hungerford* Heinemann 1977

272 'an immense series' Macmillan, Harold *War Diaries* Macmillan 1984 p.86

274 '*Après moi le déluge*' The monarch approved '*Après nous*', but this was changed to '*Après moi*'

275 'My Niggy's dead' Morris *Guy Gibson* p.178

275 'If you live' Denis Hornsey quoted Hastings *Bomber Command* p.196

275 'It was daft' Chalmers quoted Arthur p.278

275 'Can you imagine me' www.bombercommandmuseum. ca/taerum2/html

277 'For years we have' quoted Webster, Sir Charles & Frankland, Noble *The Strategic Air Offensive Against Germany 1939–45* four volumes, HMSO 1961 Vol. II p.66

277 'I have been trying' UKNA AIR4/595

278 'the proper venue' UKNA AIR14/840

278 'the battle of the Ruhr is now' UKNA AIR25/119

279 'the experience built up' Potsdam history Vol. VII p.384

280 'Bomber Command had stopped' Tooze, Adam *The Wages of Destruction: The Making and Breaking of the Nazi Economy* Penguin Books 2006 p.602

280 'The effects of this' Webster & Frankland Vol. II p.292

280 'the half-hearted nature' Fest, Joachim *Speer: The Final Verdict* Weidenfeld & Nicolson 2001 p.165

281 'Any operation of war' AI Harris 1978

282 'the decrypts showed' Hinsley et al. *British Intelligence in the Second World War* HMSO 1981 Vol. II p.673

282 'The general impression is' UKNA HW16/9

282 'the [Möhne] damage did not touch' UKNA FO837/17 MEW Intelligence weekly Report 72A 3.7.43

283 'a spectacular feat' Howard, Michael *Grand Strategy August 1942–September 1943* [British Official History series] HMSO 1970 Vol. IV p.316

284 'High Wycombe assessed' For a fuller discussion of High Wycombe's techniques of damage assessment see the author's *Bomber Command*

284 'We can wreck Berlin' CCh College Harris in Portal Papers File 9

285 'His personality was not' Dyson p.29

285 'Berlin won' AI Cochrane 1978, quoted *Bomber Command* p.268

285 'The strength of the German defences' ibid. p.268

287 'an immense sense' Moorehead *Eclipse* Granta 2000 p.221

288 'I wanted it to be' WSC in Montague Brown p.201

10: Landings

289 'unsustainable' IWM Sound Archive Shannon lecture 16256/2/1

290 'Wallis received massive support' Edgerton, David *Britain's War Machine* Allen Lane 2011 p.239

291 'I stood by [Les]' Kellow p.22

293 'in full and final' UKNA AVIA53/627

294 'Did he suffer?' Humphries p.61

294 'Sutherland, who had' Sutherland *Times* obituary 24.1.19

295 'He is a perfect poppet' Churchill, Winston S. *The Churchill Documents* Vol. XVIII 1943, ed. Martin Gilbert & Larry Arne Hillsdale College Press 2015 p.1472

296 'In spite of the fact' Pim in ibid. p.2192

296 'Of course, when' Morris *Gibson* p.216

297 'after our experience' Harris quoted ibid. p.195

297 'Mickey Rooney' ibid. p.314

298 'why had two great' Gibson *ECA* p.30

298 'I'm not a highbrow' Gibson in transcript *Desert Island Discs* broadcast 19.2.44

301 'I do feel that Guy' Eve Gibson letter in St Edward's school archive

302 'When the film *The Dam Busters*' Information from Robert Owen, told to him by Richard Todd

303 'Someone else said' Gibson *ECA* p.297

304 'It is impossible' Howard to the author in 2019

305 'The cost of' Hastings *Bomber Command* p.352

305 'There existed throughout' Overy, Richard *The Bombing War* Allen Lane 2013 p.633

306 'Secrecy pervaded' Dyson p.29

306 'The head people' IWM Sound Archive Transcript Sumpter 8177/3/2 p.44

306 'he did not doubt' Nuffield College, Tizard in Cherwell Archive F155/14 15.4.42
307 'I suddenly see' Danziger, Danny *Eton Voices* Viking 1988 p.66

Bibliography

Articles, pamphlets and websites

aircrewremembered.com/byers-vernon html

Air Historical Branch Monograph *Armament* Vol. I 1952

— Translation No.VII/36 The Westphalian President's report to the Minister of Home Affairs 24.6.43

Blank, Ralf *The night of 16/17 May 1943* Westfalische Geschichte internet portal

https://bregroup.com/about-us/our-history/dambusters/ Eduard-Joroszewski-und-das-Lager-Talsperre.pdf [an analysis of the dead in the Neheim slave labour camp]

Gibson, Guy 'Cracking the German Dams', *Atlantic Monthly* December 1943 pp.45–50 [ghosted]

— *Sunday Express* 5.12.43 'How We Smashed the German Dams' [ghosted and derived from the above]

Grantham Journal 'Dambusters Revisited' May 1993

Harvey, Elizabeth 'Last Resort or Key Resource? Women workers from the Nazi-occupied Soviet Territories, the Reich Labour Administration and the German War' *Transactions of the Royal Historical Society* pp.149–73, 2016

Morris, Richard ed. *Breaching the German Dams: Flying into History* RAF Museum 2008

— (as author) *Operation Chastise*, Studies in Contemporary and Historical Archaeology 5, BAR International 2009

— 'The Aircrew Who Flew the Dams Raid' unpublished paper 2018

— 'The Dams Raid' unpublished paper 2018

Owen, Robert *Considered Policy or Haphazard Evolution: No.617 Squadron RAF 1943–45* doctoral thesis for University of Huddersfield 2014

— 'Attacking the Sorpe' memorandum to MH 16.10.18

Page, Bruce *Sunday Times* magazine 28.5.72 'How the Dambusters' Courage was Wasted'

Royal Aeronautical Society conference 'The Dams Raid 75 Years On' London 17.5.18

RAF Historical Society 'Symposium on the Strategic Bomber Offensive 1939–45' 1993

https://www.schiebener.net/wordpress/wp-content/uploads/2018/02/33

Westfalische Geshichte Westphalian history internet portal

Wood, D. ed. *Reaping the Whirlwind* Bracknell Paper Number 4A

Books

Alanbrooke, Lord *War Diaries 1939–45*, ed. Danchev & Todman, Weidenfeld & Nicolson 2001

Arthur, Max *Dambusters: A Landmark Oral History* Virgin Books 2009

Bateman, Alex *No.617 'Dambusters' Squadron* Osprey 2009

Birrell, Dave *Big Joe McCarthy* Wingleader 2012

Bishop, Patrick *Bomber Boys* Harper Press 2007

— *Air Force Blue* William Collins 2017

Brickhill, Paul *The Dam Busters* Pan 1951

Brown, Anthony Montague *Long Sunset: The Memoirs of Churchill's Last Private Secretary* Cassell 1995

Churchill, Winston S. *The Churchill Documents* Vol. XVIII 1943, ed. Martin Gilbert & Larry Arne, Hillsdale College Press 2015

Cooper, Alan *The Dambusters Squadron* Arms & Armour 1993

Driesschen, Jan van den *We Will Remember Them* Erskine Press, Quidenham, Norfolk 2004

Dyson, Freeman *Disturbing the Universe* Harper & Row 1979

Echternkamp, Jörg ed. *Germany and the Second World War Col.IX/1 German Wartime Society 1939–45: Politicization, Disintegration and the Struggle for Survival* Research Institute for Military History Potsdam, tr. Clarendon Press 2008

Edgerton, David *Britain's War Machine* Allen Lane 2011

Elmes, Jenny *M-Mother: Dambuster John 'Hoppy' Hopgood* The History Press 2015

Euler, Helmuth *The Dambuster Raid: A German View* Pen & Sword 2015

Fest, Joachim *Speer: The Final Verdict* Weidenfeld & Nicolson 2001

Flower, Stephen *The Dam Busters: An Operational History of Barnes Wallis's Bombs* Amberley 2013

Foster, Charles *Breaking the Dams: The Story of David Maltby and His Crew* Pen & Sword 2008

— *The Complete Dambusters: The 133 Men Who Flew on the Dams Raid* History Press 2018

Frankland, Noble *History at War* DLM 1998

Friedrich, Jorg *The Fire* Columbia University Press 2006

Garfield, Simon ed. *Private Battles* Ebury Press 2006

Gibson, Guy *Enemy Coast Ahead* Michael Joseph 1946

Harris, Sir Arthur *Despatch on War Operations* Cass 1995

Hastings, Max *Bomber Command* Michael Joseph 1979

— *Warriors* HarperCollins 2005

— *Finest Years: Churchill as Warlord 1940–45* HarperCollins 2009

— *All Hell Let Loose* William Collins 2011

Higgs, Colin & Vigar, Bruce *Voices in Flight: The Dambuster Squadron* Pen & Sword 2013

Hinsley, F.H. et al. *British Intelligence in the Second World War* Vol. II HMSO 1981

Holland, James *Dam Busters* Corgi 2013

Howard, Michael *Grand Strategy August 1942–September 1943* [British Official History series] HMSO 1970

Humphries, Harry *Living with Heroes: The Dam Busters* Erskine Press 2008

Johnson, George 'Johnny' *The Last British Dambuster* Ebury Press 2014

Kellow, Robert *Paths to Freedom* Kellow 1992

Klemperer, Victor *To the Bitter End* Phoenix 2000

Koa Wing, Sandra ed. *Our Longest Days* Profile Books 2007

Lascelles, Sir Alan *King's Counsellor*, ed. Duff Hart Davis, Weidenfeld & Nicolson 2006

Middlebrook, Martin & Everitt, Chris *The Bomber Command War Diaries* Pen & Sword 2011

Morpurgo, J.E. *Barnes Wallis* Longman 1972

Morris, Richard with Colin Dobinson *Guy Gibson* Viking 1994

Murray, Iain *Bouncing Bomb Man: The Science of Sir Barnes Wallis* Haynes 2009

Nichol, John *Return of the Dambusters* William Collins 2015

Olsen, Sam *The Dambusters Vol. I: The Rise of Precision Bombing March 1943–May 1944* Leandoer & Ekholm 2010

Overy, Richard *The Bombing War* Allen Lane 2013

Owen, Robert *Henry Maudslay, Dam Buster* Fighting High Press 2014

— with Steve Darlow, Sean Feast, Arthur Thorning *Dam Busters: Failed to Return* Fighting High Press 2013

Peden, Murray *A Thousand Shall Fall* Stoddart 1988

Press, Nigel *All My Life* Lancfile Publishing 2006

Probert, Henry *Bomber Harris: His Life and Times* Greenhill 2006

Revie, Alastair *The Lost Command* Corgi 1972

Simpson, Tom *Lower Than Low* Libra Books 1995

Slessor, John *Air Power and Armies* London 1936

Speer, Albert *Inside the Third Reich* trs. Macmillan 1970

Sweetman, John *The Dambusters Raid* Arms & Armour Press 1993

Thorning, Arthur *The Dambuster Who Cracked the Dam* Pen & Sword 2008

Tooze, Adam *The Wages of Destruction: The Making and Breaking of the Nazi Economy* Penguin Books 2006

Verrier, Anthony *The Bomber Offensive* Batsford 1968

Ward, Chris; Lee, Andy; Wachtel, Andreas *Dambusters: The Definitive History of 617 Squadron at War 1933–45* Red Kite 2003

— *Dambusters: The Forging of a Legend* Pen & Sword 2009

Webster, Sir Charles & Frankland, Noble *The Strategic Air Offensive Against Germany 1939–45* four volumes, HMSO 1961

Wells, Mark *Courage in Air Warfare* Cass 1995

Wilson, Kevin *Bomber Boys: The RAF Offensive of 1943: The Ruhr, the Dambusters and Bloody Berlin* Weidenfeld & Nicolson 2006

Winterbotham, F.W. *Secret and Personal* William Kimber 1969

— *The Ultra Spy* Papermac 1981

Zuckerman, Solly *The Strategic Air War Against Germany 1939–45: Report of the British Bombing Survey Unit* London 1946

Index